Floods - Hydraulics and Hydrology

Edited by Guangwei Huang

Published in London, United Kingdom

Floods - Hydraulics and Hydrology
http://dx.doi.org/10.5772/intechopen.1001623
Edited by Guangwei Huang

Contributors
Akeem Abiodun Ademola, Blessing Funmbi Sasanya, Denis Arinabo, Guangwei Huang, Katsumi Wakigawa, Oluwafemi Oladipupo Kolajo, Shunsuke Segi, Teppei Ishiuchi, Toshiyuki Nishikori, Yusuke Kobayashi

First published in London, United Kingdom, 2025 by IntechOpen
IntechOpen is the global imprint of INTECHOPEN LIMITED, registered in England and Wales, registration number: 11086078, 167-169 Great Portland Street, London, W1W 5PF, United Kingdom

For EU product safety concerns: IN TECH d.o.o., Prolaz Marije Krucifikse Kozulić 3, 51000 Rijeka, Croatia, info@intechopen.com or visit our website at intechopen.com.

British Library Cataloguing-in-Publication Data
A catalogue record for this book is available from the British Library

Floods - Hydraulics and Hydrology
Edited by Guangwei Huang
p. cm.
Print ISBN 978-0-85466-773-4
Online ISBN 978-0-85466-772-7
eBook (PDF) ISBN 978-0-85466-774-1

Meet the editor

Dr. Huang is a professor at Sophia University in Japan and currently serves as the Dean of the Graduate School of Global Environmental Studies. He received his BSc from Fudan University in China and his MS and PhD from the University of Tokyo in Japan. Passionate about water, his research has evolved from fluid dynamics to water quality modeling and monitoring and further to integrated watershed science and management. In recent years, his main research focus has been on sustainability science from a water perspective, such as wetland conservation and wise use, flood vulnerability and resilience analysis, and environmental impact assessment. He has published more than 100 peer-reviewed papers in various academic journals. In 2019, he received the Distinguished Scientist award from the International Society for Environmental Information Sciences.

Contents

Preface

Flooding remains one of the most devastating natural hazards, affecting millions of people worldwide each year. As climate change intensifies rainfall patterns and urbanization alters watershed dynamics, the need for advanced studies on hydrology and hydraulics has never been greater. This book, *Floods – Hydraulics and Hydrology*, brings together fundamental principles, cutting-edge research, and practical applications to address the complex challenges of flood prediction, risk management and impact mitigation.

The interplay between hydrology (the study of the movement, distribution, and management of water on Earth) and hydraulics (the applied science of water flow in natural and engineered systems) is the backbone of effective flood risk assessment and mitigation. This volume bridges these two disciplines and provides a new source of information for students, researchers, engineers, and policymakers involved in water resources and disaster management.

The book is structured to provide both theoretical foundations and real-world case studies, ensuring a balanced approach to learning and application. Key topics covered in this book include flood hazard mapping, resident behavior and disaster preparedness during a flood event, flood risk in Africa, analysis of flood control facilities under climate change uncertainty, and a study of drought indices. We extend our deepest gratitude to the contributing authors, whose expertise and dedication have made this publication possible. Their diverse perspectives—spanning academic research, engineering practice, and policy development—enrich the book's content and enhance its relevance to global flood challenges.

We hope *Floods – Hydraulics and Hydrology* will serve as a valuable reference for advancing flood science and engineering, promoting innovation, and ultimately contributing to safer, more resilient communities in the face of growing water-related risks.

Guangwei Huang, PhD
Professor,
Graduate School of Global Environmental Studies,
Sophia University,
Tokyo, Japan

Chapter 1

An Important Role of Hydraulics Today: Flood Hazard Mapping

Guangwei Huang

Abstract

This chapter provides an overview of the application of hydraulics to flood hazard mapping in Japan. It explains the methods used to produce two types of flood hazard maps and highlights issues for further investigation, such as how to incorporate the process of levee collapse expansion in the vertical direction into modeling. It also provides evidence that flood hazard mapping is reliable in elucidating potential harm. Furthermore, it identifies problems with evacuation despite the provision of flood hazard maps. The purpose is to share information and serve as a call for new developments in flood hazard mapping. The information provided in this chapter can help young researchers and engineers to get a quick start in flood hazard mapping. At the same time, it can serve as an inspiration for more experienced researchers to pursue further in-depth studies.

Keywords: flood hazard map, levee breach, evacuation, road, Japan

1. Introduction

Hydraulics is a branch of applied science and engineering that is essentially concerned with the movement of water. It aims at better understanding the physical processes of water flow analytically and empirically, expressing these processes mathematically, and use this knowledge in the design of various water-related structures and systems. Fluid mechanics provides the theoretical foundation for hydraulic applications. Traditionally, the main focus of hydraulics has been the prediction of water depth and flow velocity in open channels, pipes, and rivers. A particular role is to provide river managers with short-term predictions of variations in river discharge and depth during a rainfall event for emergency response. Hydraulic methods have been well developed and extensively applied to rivers around the world.

In recent years, an important application of hydraulics has been its use in flood hazard mapping. Flood maps are a tool that communities use to know the likelihood of flooding in their area and which areas may have relatively higher inundation depths. Floods do not follow city limits or property lines. A flood hazard map allows residents to see the relationship between their properties and areas most at risk of flooding.

In Japan, 49% of the population and 75% of the real estate is concentrated in flood-prone areas. Therefore, flood hazard maps are essential for understanding flood exposure and vulnerability. It also provides information on emergency shelters and evacuation routes. Overall, it serves as a tool to support urban development planning, flood policy analysis, and the implementation of flood preparedness and management strategies. Given that flood disasters are common worldwide and affect the largest number of people [1], information sharing and communication on flood hazard mapping is essential to improve flood management on a global scale.

2. Method for flood hazard mapping

The flood hazard mapping methodology used in Japan has three components. These are described below.

2.1 Scenario setting

According to the amendment of the Flood Fighting Act (FFA) of Japan in 2015, the mapping process should consider two rainfall scenarios as inputs for inundation simulation. The first scenario is to use the conventional design rainfall, which is equivalent to the 100-year or 200-year return period, as an input to simulate flood inundation. The design rainfall is obtained by conventional rainfall-runoff modeling approaches. The second scenario is to use the maximum possible rainfall rate (worst case), which is equivalent to a 1000-year class rainfall, as input to simulate flood inundation. Based on rainfall patterns, Japan is divided into 15 regions. Using the recorded maximum rainfall in each region, a depth-area-duration (DAD) analysis of rainfall is performed to evaluate the maximum amount of rainfall over areas of different sizes within different durations [2]. **Figure 1** shows the maximum rainfall envelope for the Hokuriku region [2]. DAD curves are

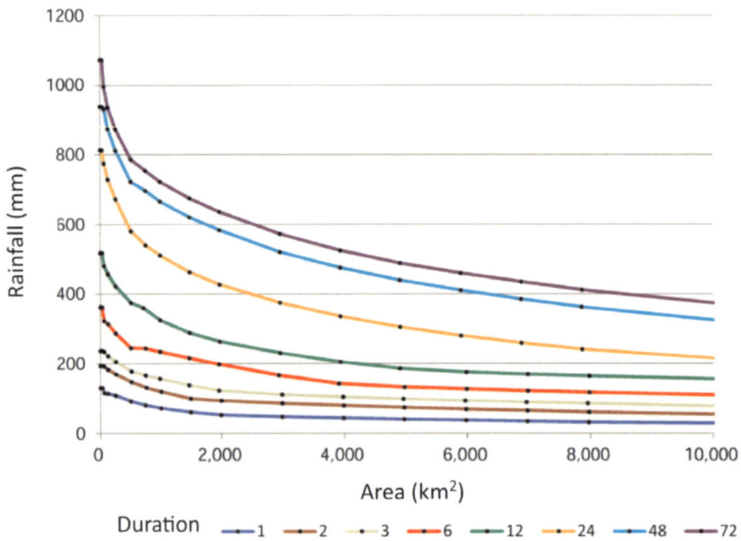

Figure 1.
DAD analysis for the Hokuriku region.

developed for durations of 1, 2, and 3-days, as shown in **Figure 1**. The analysis was performed based on isohyetal maps [3, 4], and the procedure for producing DAD curves is given below

1. Determine the day of greatest average rainfall, two consecutive days of greatest average rainfall, and move up to five consecutive days.

2. Plot a map of maximum 1-day rainfall and construct isohyets using the Inverse Distance Weighting (IDW) method; similarly prepare isohyetal maps for 2 and 3-day rainfall separately.

3. The isohyetal map, say, for maximum 1-day rainfall, is divided into zones to represent the principal rainfall centers.

The difference between the design rainfall and the maximum possible rainfall in several river basins in Japan is shown in **Table 1**. It can be seen that the difference can be more than double.

Because this estimate of the maximum possible rainfall rate is based on the area alone, it cannot account for differences in topographic conditions that affect rainfall, such as basin elevation and slope direction. In addition, the DAD analysis is derived from the maximum envelope of actual rainfall, so it does not directly account for the effects of climate change. To better account for the effects of climate change, a climate model can be used to obtain the Probable Maximum Precipitation (PMP), which is the largest accumulation of precipitation for a given duration that is meteorologically possible for an area. However, estimations of PMP are associated with large uncertainties. A study by Kunkel showed that PMP values will increase in the future due to higher atmospheric moisture content and, consequently higher moisture transport in storms [5]. Therefore, using a range of PMP values based on uncertainty evaluation, that is, the range of possible values under equivalent conditions, rather than relying only on single point estimates is necessary and more suitable for flood hazard mapping, but has been largely ignored so far.

2.2 River flood routing and the estimation of flow through a levee breach

Flood routing is a mathematical procedure for predicting the change in water depth and velocity along a waterway. The methods are well developed, and some of them, such as HEC-RAS and iRIC, are in the public domain. Therefore, this part of

Basin name	Maximum rate of rainfall possible	Design rainfall
Kiso River	527 mm in 48 h	295 mm in 48 h
Nagara River	421 mm in 12 h	243 mm in 12 h
Yodogawa River	360 mm in 24 h	261 mm in 24 h
Tado River	836 mm in 24 h	337 mm 24 h
Ijira River	430 mm in 6 h	240 mm 6 h

Table 1.
A comparison between maximum rate of rainfall possible and design rainfall in several river basins.

flood hazard mapping can be considered mature. On the other hand, model development for predicting the discharge from a levee breach remains limited. Currently, there are no widely accepted empirical equations for levee breach dimensions, lateral erosion rates, or breach development time. As a matter of fact, man-made levees can fail in a number of ways. One is overtopping [6], which consists of three stages. The first stage, called the slope erosion stage, is the flow-induced erosion of the embankment slope on the land side, forming some small scour holes on the embankment slope. The second stage, called the head-cut retreat stage, is the enlargement of the scour holes, steepening the slope and then developing a large scarp known as a head-cut. At the end of this stage, the breach cross-section was nearly washed out, and a deep gully formed. The flow in the gully then began to erode the sides of the breach, which is the third stage of erosion, defined as the breach widening stage. Another mechanism of levee failure is water infiltration into the body of a levee [7], as shown in **Figure 2**. A closely related phenomenon, also shown in **Figure 2**, is the seepage of water under a levee during a flood, which is referred to as piping. A recent work investigated the coupling between backward erosion piping and slope instability-induced levee failure [8].

At present, there is no good method that can quantitatively assess the progressive seepage and piping that can lead to a levee failure. In Japan, Homan's formula is used to estimate the flow out of the river channel due to a breach.

The actual calculation is that flood routing is performed every 200 m along the waterway, assuming that the levees would breach when the increased river water from rainfall reaches the maximum discharge that can be safely accommodated by the current levees. The results of the inundation simulations for all breach points are then overlaid to produce a map of the distribution of the deepest inundation depths in a target area. In reality, however, the probability of levee failure varies from location to location, and this difference in probability is not accounted for. In addition, the mode of levee failure is not considered in the simulations. The flow going out of the channel from a levee breach site depends on the causes of the breach, which are complicated

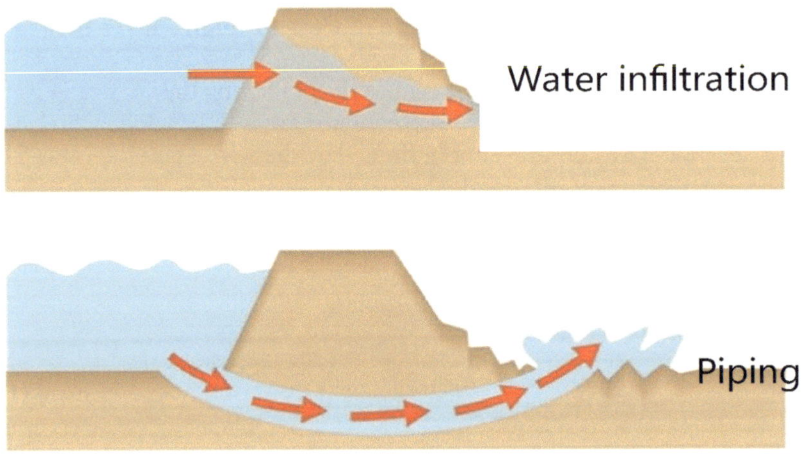

Figure 2.
Two mechanisms of levee breach.

and difficult to model the process of levee collapse. Therefore, further research is needed to overcome this shortcoming by considering levee breach mechanism in estimating the flow going out of the channel during the collapse. For the levee breach width, it is estimated using the empirical relationships between breach width and river width as follows:

$$B = 2.0 \times (\log_{10} X)^{3.8} + 77 \text{ near a river junction} \tag{1}$$

$$B = 1.6 \times (\log_{10} X)^{3.8} + 62 \text{ at other location} \tag{2}$$

where B is the width of the breach and X is the width of the main river channel. The time progression of the levee breach is assumed to have one-half of the final breach width instantly at the onset of breaching, followed by an expansion to the final breach width in 1 h. The rate of expansion of the width during this period is assumed to be constant, and it is assumed to expand in both the upstream and downstream directions. An issue is that in the vertical direction, the expansion to the bed is assumed to occur instantly. How the process of a levee collapses in the vertical direction remains an unsolved problem.

To calculate the outflow from a levee breach, the first step is to use Homan's front-overflow formula as given below.

For complete overflow ($h_2/h_1 < 2/3$) and submerged overflow ($h_2/h_1 \geq 2/3$),

$$Q_0 = \begin{cases} m h_1 B \sqrt{2 g h_1} & \dfrac{h_2}{h_1} < \dfrac{2}{3} \\ \dfrac{3\sqrt{3}}{2} m h_2 B \sqrt{2g(h_1 - h_2)} & \dfrac{h_2}{h_1} \geq \dfrac{2}{3} \end{cases} \tag{3}$$

where Q_0 is flow discharge through a weir, h_1 and h_2 are the water depth on the two sides of the weir measured from the river bed. B is the width of the breach.

Then, Honma's side-overflow formula is used to estimate the inundation discharge as given by:

For $I > 1/1580$,

$$\frac{Q}{Q_0} = \left(0.14 + 0.19 \times \log_{10}\left(\frac{1}{I}\right) \times \cos\left(48 - 15 \times \log_{10}\left(\frac{1}{I}\right)\right)\right) \tag{4}$$

For $1/1580 \geq I > 1/33{,}600$,

$$\frac{Q}{Q_0} = \left(0.14 + 0.19 \times \log_{10}\left(\frac{1}{I}\right)\right) \tag{5}$$

For $1/33{,}600 \geq I$,

$$\frac{Q}{Q_0} = 1 \tag{6}$$

where Q is the inundation flow, Q_0 is flow volume calculated by Honma's formula (3), I is the bed slope of a river.

2.3 2-D inundation simulation

The following shallow water equations (SWE) are used for the simulation of flood inundation [9].

$$\gamma\frac{\partial Q_x}{\partial t}+\frac{\partial}{\partial x}\left(\gamma\frac{Q_x^2}{h}\right)+\frac{\partial}{\partial y}\left(\gamma\frac{Q_xQ_y}{h}\right)+g\gamma h\frac{\partial(h+z_b)}{\partial x}+g\gamma n^2\frac{Q_x\sqrt{Q_x^2+Q_y^2}}{h^{7/3}}$$
$$+\frac{1}{2}C_D'(1-\gamma)\frac{Q_x\sqrt{Q_x^2+Q_y^2}}{h}=0\gamma\frac{\partial Q_y}{\partial t}+\frac{\partial}{\partial x}\left(\gamma\frac{Q_xQ_y}{h}\right)+\frac{\partial}{\partial y}\left(\gamma\frac{Q_y^2}{h}\right)+g\gamma h\frac{\partial(h+z_b)}{\partial y}$$
$$+g\gamma n^2\frac{Q_y\sqrt{Q_x^2+Q_y^2}}{h^{7/3}}+\frac{1}{2}C_D'(1-\gamma)\frac{Q_y\sqrt{Q_x^2+Q_y^2}}{h}=0\frac{\partial h}{\partial t}+\frac{\partial(\gamma Q_x)}{\partial x}+\frac{\partial\left(\gamma Q_y\right)}{\partial y}=q \tag{7}$$

where, Q_x, Q_y are the discharges per unit width in x and y directions, h is the water depth, z_b is the bed elevation, γ is the porosity, q is the rainfall, outflow from the sewer, etc., n is the roughness coefficient according to the land use, $C_D{}'$ the drag coefficient for buildings. The spatial resolution for inundation simulation is usually 250 m × 250 m or 50 m × 50 m.

In a flood hazard map, inundation depth is divided into either six (standard) or eight (detailed) categories marked by different colors, as shown in **Figure 3**. Such a division may be subject to change when applied to another country, as it is based on the average height of Japanese people and the standard two-story house design. The color schemes are determined in accordance with ISO and other standards, with consideration for people having color blindness.

Figure 4 shows the difference in the flood hazard map between the design rainfall-based and the worst case for a reach of the Chikuma River. It can be clearly seen that the inundation area is expanded and the depth is deepened in the worst-case scenario.

Even in areas where the depth of flooding is shallow, prolonged flooding will hinder the restoration of lifelines (electricity, gas, and water) and affect daily life. Therefore, the duration of inundation is another important indicator in addition to

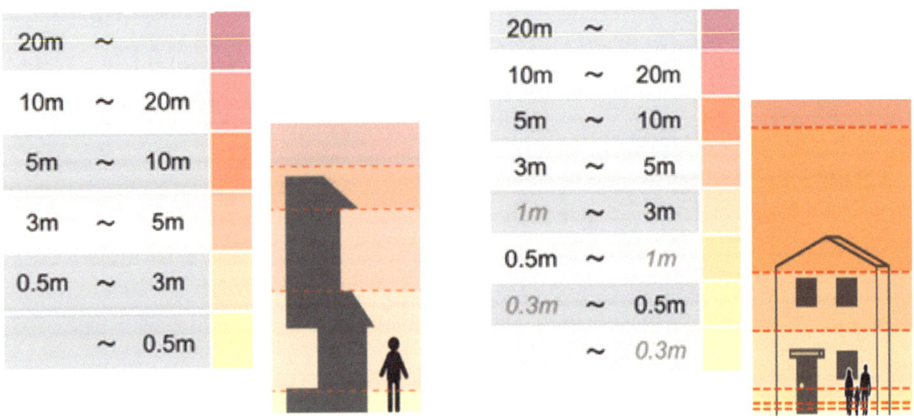

(a) Standard (b) Detailed

Figure 3.
Color bar used in flood hazard map.

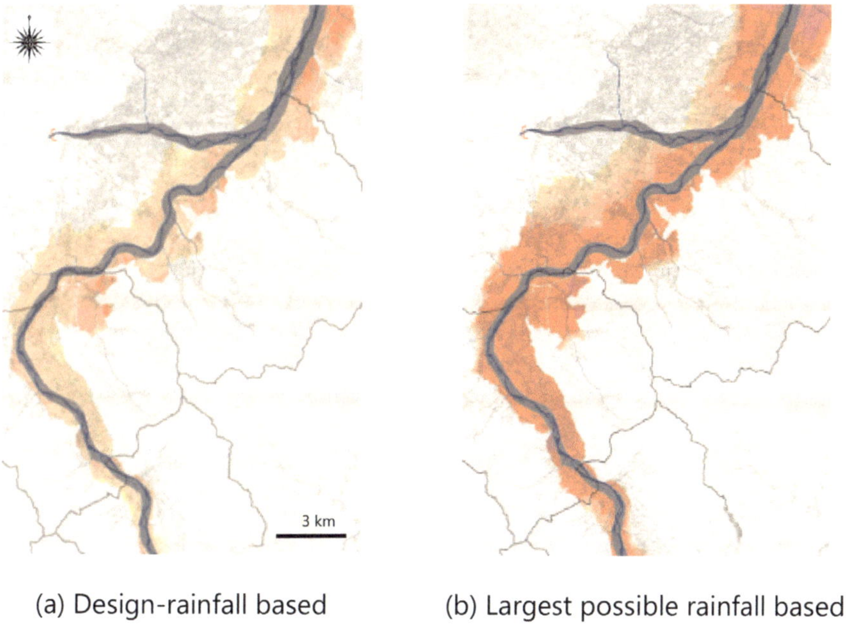

Figure 4.
Comparison of two types of flood hazard map (source: Chikuma River Office).

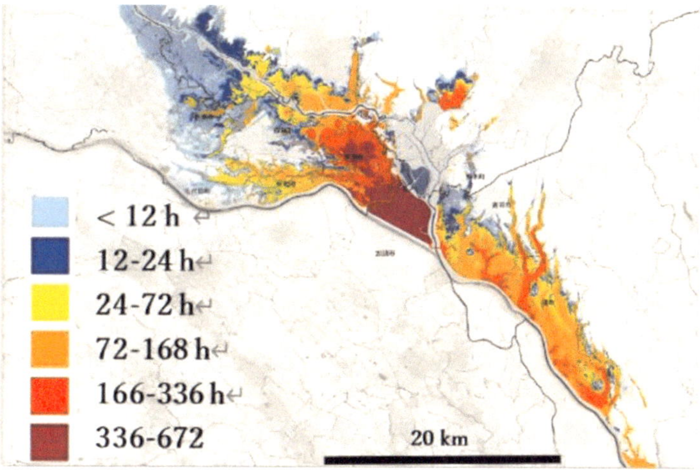

Figure 5.
An illustrative example of inundation duration.

inundation depth. In Japan's flood hazard map, the duration of inundation is classified into three ranges, with areas under 12 h (light blue), 12 h to 1 day (blue), 1 day to 3 days (yellow), and 3 days to 1 week (orange), 1 week to 2 weeks (red) and 2 weeks to a month (dark red) as shown in **Figure 5**.

At present, 98% of municipalities in Japan have prepared flood hazard maps for the design storm and made them available to residents, while 83% of municipalities have mapped flood hazard areas for the largest-scale rainfall. Although flood hazard maps are well prepared, there is a need to increase people's awareness of flood hazard maps. It has been found that the people who have not seen the flood hazard map spend more than 1 h to evacuate than the people who have seen the flood hazard map in advance [10]. Therefore, it is important to increase individual awareness of flood hazard areas and evacuation options. In particular, people who move to a new city are often unaware of the flood risk in the region, so their awareness should be raised in one way or another. Today, real estate agents are responsible for providing flood risk information to their clients prior to the transaction.

Validation of flood hazard maps is difficult because it is not made for a single actual storm event. Nevertheless, there are examples of good matches between predicted flooding areas and actual inundated areas. In 2018, along the Oda River system, a tributary of the Takahashi River, river water ran out of the channel at multiple locations with different mechanisms from levee breach and partial levee collapse to overflow without levee damage, resulting in a major disaster that submerged the entire Mabi Town area. As shown in **Figures 6** and 7, the actual

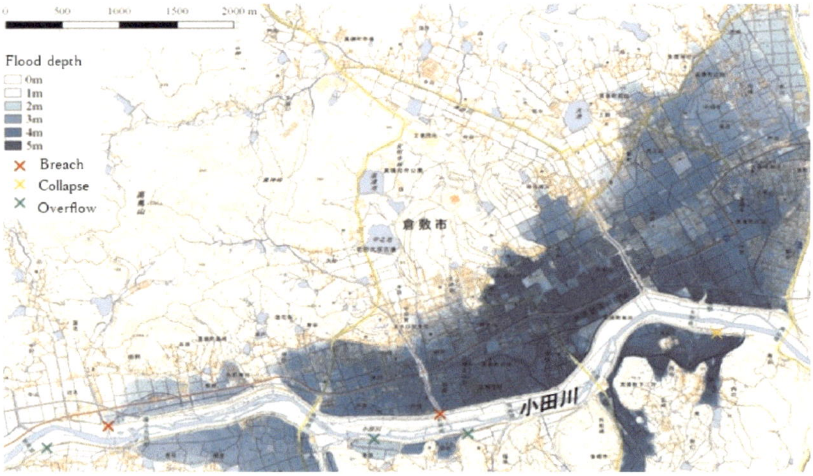

Figure 6.
The inundated areas in 2018 (source: [11]).

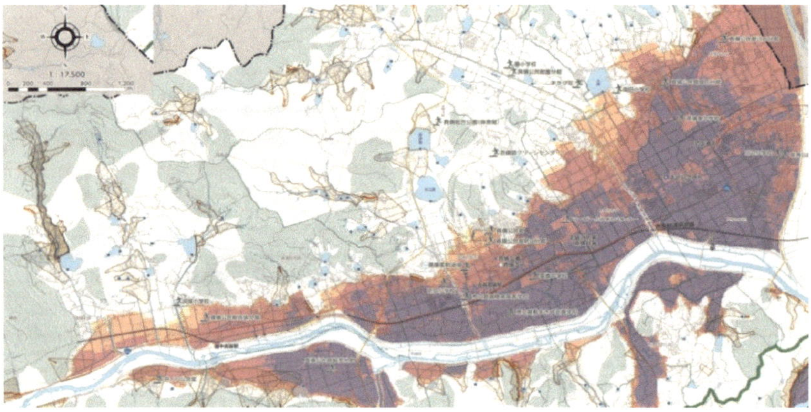

Figure 7.
The design rainfall-based flood hazard map for Mabi Town (source: [12]).

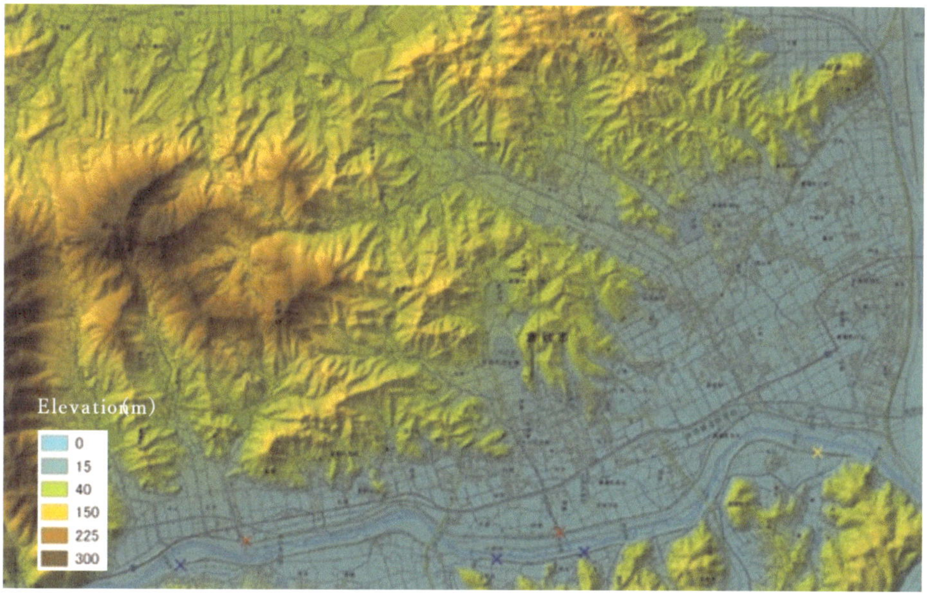

Figure 8.
Land surface elevations of Mabi Town.

inundation areas coincided with the potential flood zones highlighted by the flood hazard map. **Figure 8** shows that both the actual and predicted inundation areas match the lowlands of Mabi Town. This case suggested a hypothesis that both the actual and predicted inundation areas are more affected by topography than the location and mechanism of the levee breach under certain circumstances. Further studies should be pursued to test this new hypothesis.

3. Problems with evacuation

A fundamental question that needs to be addressed is why we need a flood hazard map. A commonly accepted answer is to support policy analysis, assist land use management, facilitate public participation in flood risk management, and support emergency response. For effective evacuation, the right information about shelters and evacuation routes should be provided to residents. However, many shelters are in fact, located in high-risk areas and may be flooded and inoperable.

In 2019, Typhoon No. 19 brought about record-breaking rainfall across East Japan from 12 to 13 October 2019, causing widespread flooding in many areas, including Tokyo, Chiba, and Nagano prefectures, and affected more than 7 million people.

Figure 9 shows that shelters in Naganuma Ward, Nagano City, that functioned and malfunctioned during a flood disaster in 2019 [13]. The yellow circles indicate shelters that are close to the river and failed to function as shelter according to the author's investigation. The blue triangles show shelters that were actually used during the disaster. It also showed that there were welfare facilities at unsuitable locations.

Figure 10 shows the flood hazard map of Edogawa Ward in Tokyo and the suggested evacuation routes as indicated by arrows [14]. Edogawa Ward is a flood-prone area because 70% of the land is below sea level. As a result, evacuation routes pass through places that may be flooded seriously, making them unsafe if evacuation

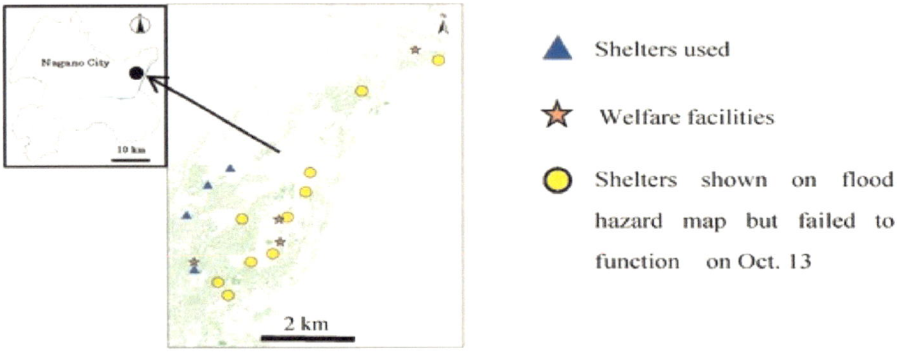

Figure 9.
An example of malfunctioning shelters in a flood disaster (source: [13]).

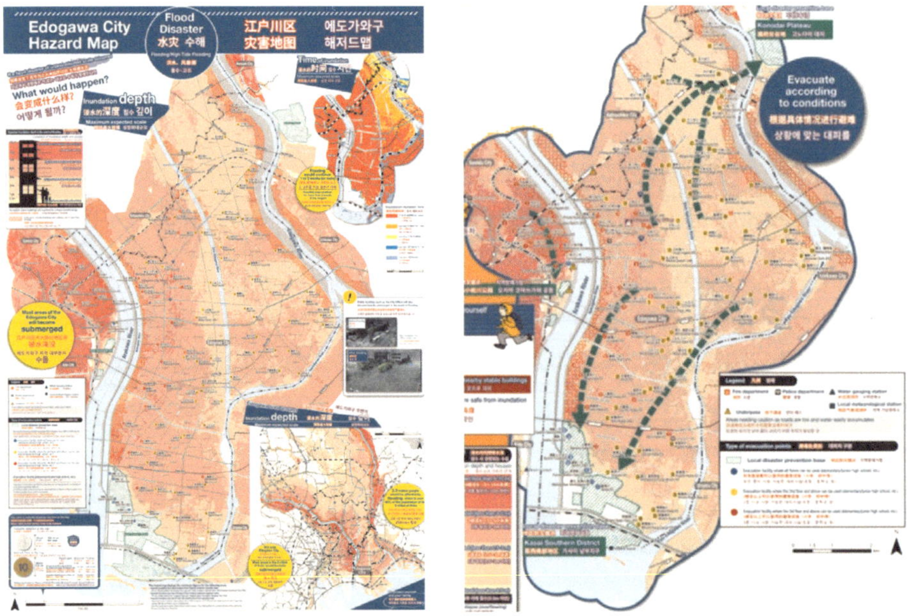

Figure 10.
Flood hazard map of Edogawa Ward, Tokyo, and the suggested evacuation routes (source: [14]).

takes place after flooding has started. **Figure 11** shows that all roads surrounding a designated flood evacuation shelter were flooded at least once during 2010–2016 [14]. This indicates that after a hazard map is prepared in a region, an evacuation route safety assessment should be conducted.

Roads are designed and built to move people and goods. Road safety refers to the methods and measures used to prevent death and serious injury to road users. However, concerns about road safety during evacuation have never been incorporated into the principle of road design. In Japan, tragedies have occurred during flood evacuations, suggesting that designated evacuation routes may be life-threatening due to insufficient consideration of road flooding.

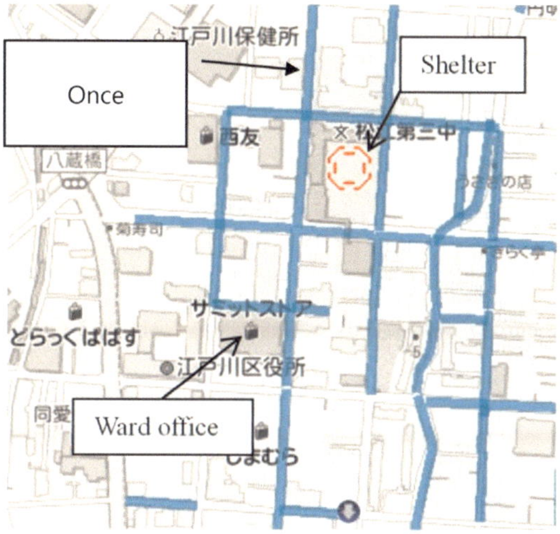

Figure 11.
A shelter without safe access roads (source: [15]).

4. Conclusion

Hydraulic engineering is a traditional discipline. One of its tasks today is the simulation of flood inundation for the preparation of flood hazard maps. Although this application is well advanced, the following issues remain to be addressed.

1. Modeling the flow from a river channel toward a levee breach, taking into account the levee breach mechanisms and the probability of failure. However, there are cases indicating that the inundation area may be more controlled by topography than the breach mechanism.

2. Evaluating the appropriateness of flood emergency shelters based on flood hazard maps and working out solutions.

3. Develop methods for evaluating the safety of evacuation routes, which may require floe velocity information in addition to inundation depth. Innovation is needed since flow velocity simulation is more sensitive to modeling details.

Author details

Guangwei Huang
Graduate School of Global Environmental Studies, Sophia University, Japan

*Address all correspondence to: huanggwx@sophia.ac.jp

References

[1] Membele GM, Naidu M, Mutanga O. Examining flood vulnerability mapping approaches in developing countries: A scoping review. International Journal of Disaster Risk Reduction. 2021:102766

[2] Water Management and Land Conservation Bureau, Ministry of Land, Infrastructure, Transport and Tourism. (2015). Methodology for setting the assumed maximum external force for the preparation of inundation map. Available form: https://www.mlit.go.jp/common/001097591.pdf. (in Japanese). [Accessed: March 10, 2025]

[3] World Meteorological Organization (WMO), 1961. Manual of depth-area-duration analysis of storm precipitation. No. 237. Available form: https://library.wmo.int/viewer/41892/download?file=wmo_237.pdf&type=pdf&navigator=1. [Accessed: March 10, 2025]

[4] Othman N et al. Investigation of Depth-Area-Duration (DAD) Curves for Kuantan River Basin (KRB). In: IOP Conference Series: Earth Environmental Science 1296 012016. 2024

[5] Kunkel K, Karl TR, Easterling D, et al. Probable maximum precipitation and climate change. Geophysical Research Letters. 2013;**40**:1402-1408. DOI: 10.1002/grl.50334

[6] Flynn S, Vahedifard F, Schaaf D. A dataset of levee overtopping incidents. Geo-Extreme 2021. 2021:99-108. DOI: 10.1061/9780784483701.010

[7] Chen S, Zhong Q, Shen G. Numerical modeling of earthen dam breach due to piping failure. Water Science and Engineering. 2019;**12**(3):169-178

[8] Rahimi M, Shafieezadeh A. Coupled backward erosion piping and slope instability performance model for levees. Transportation Geotechnics. 2020;**24**: 100394. DOI: 10.1016/j.trgeo.2020.100394

[9] National Institute for Land and Infrastructure Management. Manual for Flood Hazard Mapping Manual. 4th ed2015. Available form: https://www.mlit.go.jp/river/shishin_guideline/pdf/manual_kouzuishinsui_1507.pdf. (in Japanese). [Accessed: March 10, 2025]

[10] Katada T. Flood hazard mapping from the perspective of disaster risk communication. Fire and Disaster Prevention Science. No. 128. 2017:18-22

[11] Geospatial Information Authority of Japan. Available form: https://www.gsi.go.jp/BOUSAI/H30.taihuu7gou.html

[12] Kurashiki City. Available form: https://www.city.kurashiki.okayama.jp/1870.htm. [Accessed: Feb 1, 2025]

[13] Fan J, Huang G. Evaluation of flood risk Management in Japan through a recent case. Sustainability. 2020;**12**:5357. DOI: 10.3390/su12135357

[14] Edogawa City. Available form: https://www.city.edogawa.tokyo.jp/e007/bosaianzen/bosai/kanrenmap/n_hazardmap.html. [Accessed: Feb 1, 2025]

[15] Huang G. Enhancing dialogue between flood risk management and road engineering sectors for flood risk reduction. Sustainability. 2018;**10**:1773. DOI: 10.3390/su10061773

Chapter 2

Evacuation Behaviors and Disaster Preparedness of Local Residents during River Flooding: The Case of Typhoon 19 of 2019 in Japan

Teppei Ishiuchi

Abstract

Japan is known as a disaster-prone country and has experienced frequent heavy rainfall disasters in recent years. Typhoon No. 19 in 2019 caused extensive damage in many parts of Japan. Therefore, this chapter describes the behaviors of residents in areas where river flooding occurred and their usual awareness of disaster preparedness at the time. In particular, the situation of residents who evacuated or did not evacuate and their usual awareness of disaster prevention are described. In addition, the results of a survey on disaster preparedness will be described by conducting interviews with residents living in a landslide hazard warning zone, which is a particularly dangerous area in the event of a torrential rain disaster.

Keywords: floods, river overflows, landslides, evacuation behavior, disaster awareness

1. Introduction

Approximately 70% of Japan's land is covered with forests, meaning there is little flat land with significant differences in elevation. As a result, the entire lengths of rivers are short compared to their counterparts in Europe and the America, causing more rapid rivers in Japan due to the same elevation difference over a shorter distance [1]. Japan, with its beautiful natural environment that is the envy of the world, is vulnerable to disasters due to its topography, geology, weather, and other characteristics. The country faces exceptionally harsh natural conditions. In addition, Japan is generally located in the temperate zone, with four distinct seasons: spring, summer, fall, and winter. In addition to steep rivers, the population and assets are concentrated in low-lying areas at river mouths, making the country vulnerable to natural disasters [2]. However, when considering urban disaster prevention, flood phenomena and management cannot be simply addressed by "not using land in that area because flooding is anticipated." Given the current situation where populations and assets are concentrated in low-lying areas of these estuaries, it is necessary to view events assuming that people will continue to live there, including areas that are expected to be inundated [3]. Japan is a disaster-prone country in the middle of many typhoon' paths. As such, damage related to wind and flood damage, especially typhoons and torrential rains, occur frequently. Typhoons not only cause heavy rainfall, flooding, and storm surges but also cause slope failures and landslides that significantly impact residents' lives.

In recent years, natural disasters have occurred every year on our land, such as the torrential rains in northern Kyushu in July 2017, the Heavy rain event of July 2018, Typhoon No. 21 in 2018, Typhoon Faxai in 2019, Typhoon Hagibis in 2019, the torrential rains in July 2020, and Typhoon No. 10 in 2020. Regarding precipitation, Typhoon Hagibis 2019 brought the highest 24-hour precipitation ever recorded at 103 locations nationwide. The frequency of short-duration downpours has increased by about 1.4 times over the past 30 to 40 years. In the case of rivers, the number of waterways exceeding the flooding danger level is increasing, and there are concerns that water-related disasters will become more frequent and severe in the future [4]. Of the above, the total number of fatalities and missing persons in Typhoon Hagibis in 2019, which was a historic rainstorm, exceeded 100. This shows that reducing the number of affected people by floods has become increasingly important in recent years.

This study focuses on Typhoon Hagibis in 2019 and aims to clarify the damage, evacuation behavior, and disaster prevention awareness of the residents in the inundated area, as well as the awareness of residents that they live in sediment disaster hazard zones.

2. Legal framework for disaster countermeasures in Japan

Disasters have increased rapidly in Japan after World War II. The causes include Japan's natural environment, which leads to frequent damage caused by earthquakes, typhoons, and human factors. Specifically, it has been pointed out that while deforestation progressed during the long war, afforestation activities were neglected, thus decreasing the flood mitigation function of forests. Flood control projects that had been proceeding in an orderly fashion before the war were temporarily suspended and various legal regulations were relaxed in the postwar period [5]. In September 1959, the Isewan Typhoon (930 hPa, 5098 fatalities or missing persons) triggered the enactment of the Basic Act on Disaster Management in 1961 to ensure comprehensive and systematic disaster countermeasures and to establish a system to cope with large-scale disasters over a wide area. The purpose of the Act is to protect our land and citizens' lives, bodies, and property from disasters and to contribute to maintaining social order and ensuring public welfare. To this end, it stipulates clarification of responsibilities related to disaster management, organization pertaining to disaster management, disaster management plans, promotion of disaster management measures, fiscal and financial measures, states of emergency during disasters, and other factors [6].

While there are various disaster preparedness measures depending on the stage of the disaster, the four phases of disaster response proposed by the U.S. Federal Emergency Management Agency (FEMA) are well-known. The four stages are mitigation, preparedness, response, and recovery. It is a cyclical process that returns to prevention after recovery is complete [7, 8]. Based on the Basic Act on Disaster Management, the National Disaster Management Council prepared the Basic Disaster Management Plan as the foundation of Japan's disaster countermeasures. This plan provides basic policies for establishing a disaster management system, promoting disaster management projects, speedy and appropriate disaster recovery, and promoting science, technology, and research related to disaster prevention. The plan includes various measures for each disaster type in the mitigation, preparedness, response, and recovery stages described above.

In addition, the Basic Act on Disaster Management requires the heads of Designated Administrative Organizations, Local Administrative Agencies, Designated Public Corporations, and Local Public Corporations to formulate a regional disaster prevention plan for the area in question based on the aforementioned Basic Disaster Management Plan. However, since its enactment in 1961, the Basic Act on Disaster Management has been revised in response to repeated disasters such as the Great Hanshin-Awaji Earthquake and the Great East Japan Earthquake [9].

In particular, learning from the lessons of Typhoon Hagibis in 2019, which caused expansive and extensive damage mainly in the Tohoku and Kanto Koshinetsu regions, evacuation measures were strengthened in response to increasingly severe and frequent torrential rain disasters. Specifically, the measures include (1) the unification of evacuation recommendations and evacuation instruction, (2) Preparation of individual evacuation plans, (3) Establishment of a National Disaster Management Headquarters at a stage when a disaster is likely to occur, etc., and (4) strengthening of the government's disaster risk management measures [10].

As such, this study focuses on Typhoon Hagibis in 2019, which significantly impacted disaster management in Japan.

3. Damage caused by Typhoon No. 19 in 2019

Typhoon Hagibis, which originated off the coast of Minamitorishima island on October 6, moved westward over the Mariana Islands and temporarily developed into a large and ferocious typhoon, then gradually changed its course to the north, moved northward south of Japan, and made landfall on the Izu Peninsula with large and strong force just before 7:00 p.m. on October 12. It then passed over the Kanto region and turned into an extratropical cyclone east of Japan at 12:00 p.m. on October 13. With the approach and passage of Typhoon Hagibis, heavy rain, storms, high waves, and storm surges were observed over a wide area.

As for rain, total precipitation from the 10th to the 13th reached 1000 mm in Hakone, Kanagawa Prefecture, and exceeded 500 mm at 17 locations, mainly in eastern Japan. In particular, many locations in Shizuoka and Niigata Prefectures, the Kanto Koshin region, and the Tohoku region experienced record-breaking rainfall, with the highest 3-, 6-, 12-, and 24-hour precipitation amounts ever recorded. Heavy rain emergency warnings were issued for Shizuoka, Kanagawa, Tokyo, Saitama, Gunma, Yamanashi, Nagano, Ibaraki, Tochigi, Niigata, Fukushima, Miyagi, and Iwate prefectures in sequence from 3:30 p.m. on October 12. After the maximum warning was issued across these 13 prefectures, all warnings were lifted by 8:40 a.m. on October 13.

The heavy rains caused rivers to overflow over a wide area, causing landslide disasters and inundation damage. These heavy rains and storms caused human casualties, damage to homes, and damage to lifelines such as electricity, water, roads, and railroad facilities. They also caused transportation disruptions, including the suspension of air and rail services [11].

The target areas for this study are as follows. For river flooding, the Tairashimo-hirakubo District in Iwaki City, Fukushima Prefecture, and the Funaoka District in Shibata Town, Miyagi Prefecture, which were mentioned above, were studied. The Matsuyama and Naruko Districts in Osaki City, Miyagi Prefecture, which have suffered the most landslides in Miyagi Prefecture on record, were studied for landslide disasters. Note that the Matsuyama and Naruko Districts in Osaki City sustained no damage from Typhoon Hagibis. **Figure 1** shows the location and river names of the surveyed areas. **Table 1** shows the two locations in Osaki City; the study area focused on landslide disasters and the damage to Iwaki City and Shibata Town caused by Typhoon Hagibis in 2019 [12, 13]. **Figure 2** shows an image of the Tairashimohirakubo District in Iwaki City at the time of inundation.

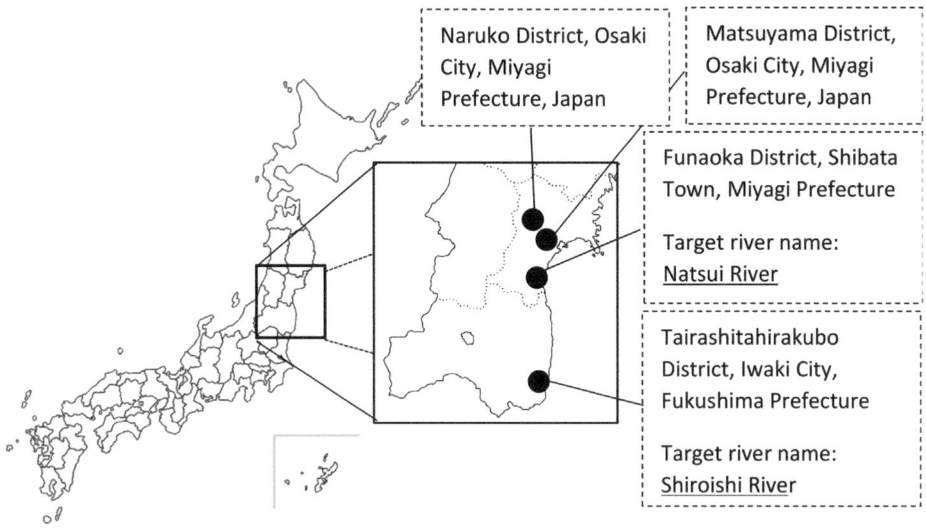

Figure 1.
Location of the study area and names of the target rivers.

		Iwaki ctiy	Shibata Town
Overview of the city or town	population	318,946 persons.	36,560 persons.
	area	1232.26 km^2	54.03 km^2
Number of personal damage	dead person	14 ※Of which, disaster-related deaths. 6 persons	0
	unidentified person	0	0
Number of the injured	serious illness	0	0
	minor injury	31	※ Traffic jam rear-end collision
Number of housing units damaged.	complete destruction	103	1
	Large-scale semi-destruction	759	4
	semi-destruction	3001	280
	Only partially damaged (semi-substantial destruction)	123	271
	Only partially damaged (less than 10%)	1123	602
Number of damages	inundation above floor level	4034	625
	inundation under floor level	951	533

Table 1.
Damage in Iwaki City and Shibata Town caused by Typhoon No. 19 in 2019.

Figure 2.
Flooding in the Tairashitahirakubo District of Iwaki City.

4. Evacuation behavior and disaster awareness during Typhoon No. 19 in 2019

4.1 Inundation depth survey in the field

4.1.1 Overview of inundation depth survey

Field surveys were conducted to determine inundation depths in the Tairashimohirakubo District of Iwaki City, Fukushima Prefecture, and Shibata Town, Miyagi Prefecture—the target areas of this study. **Figure 3** shows an overview of the

	TairaShimohirakubo District, Iwaki City	Funaoka District, Shibata Town
date of survey	December 14, 2019; January 5, 2020; January 13, 2020	December 15-18, 2020
Number of survey points	11 points	6 points
Observation scenery		
aerial photograph		

Figure 3.
Overview of flood scar survey on-site.

inundation depth survey, including a view of the survey and an aerial photograph of the target area [14]. In the field survey in Shibata Town, in particular, the number of observation points was small due to the detection of inundation scars because approximately 1 year had passed since the disaster.

4.1.2 Flood hazard map and inundation depth study results and discussion

Figure 4 shows the results of flood hazard maps [15] and inundation depth studies. The legend of the hazard map shown in **Figure 4** includes the following guidelines: 0.0 to 0.5 m is knee-deep in water for adults; 0.5 to 3.0 m is up to the ceiling of the first floor; 3.0 to 5.0 m is up to the second floor, and 5.0 to 10 m is up to the roof of the second floor.

Figure 3 shows that the land use in the Tairashimohirakubo District of Iwaki City has a high percentage of residential land, but the maximum expected inundation is 5 to 10 m. In the case of Typhoon Hagibis in 2019, which is the subject of this study, the deepest point of inundation was 2.2 m, as shown in **Figure 4**. It was confirmed that vertical evacuation to a height of about two stories would not result in human casualties.

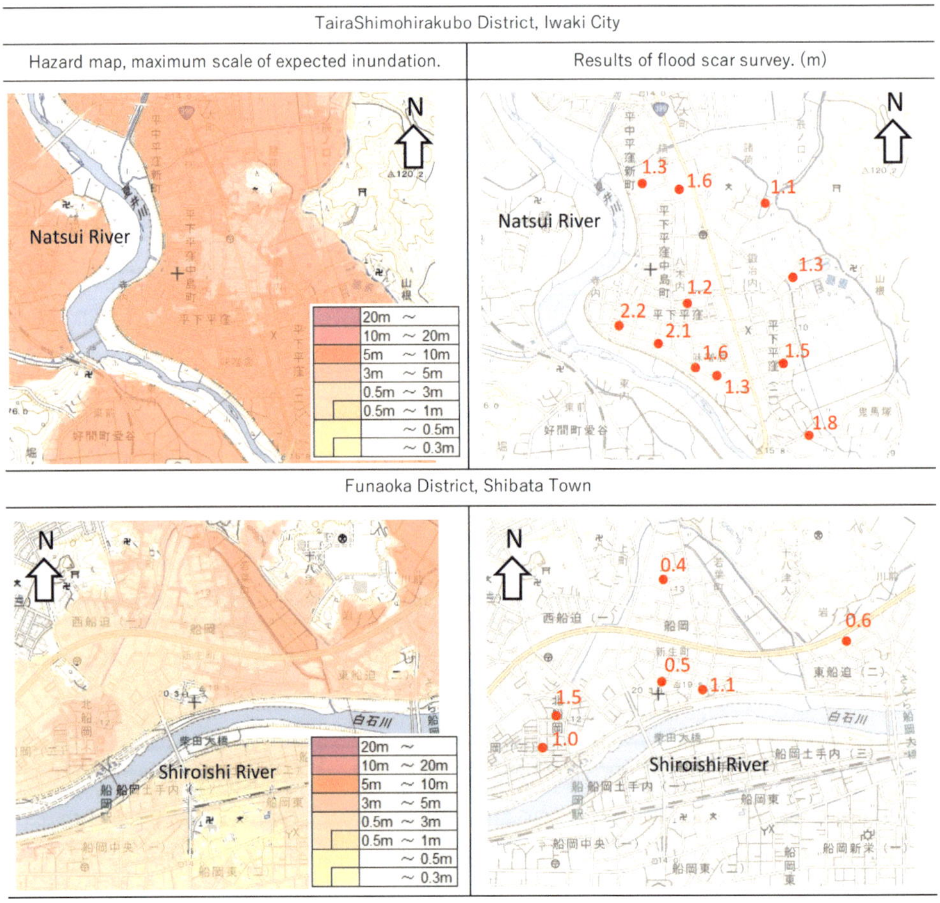

Figure 4.
Hazard map and flood scar survey results.

In addition, since inundation depths of 1 m or more were observed in a wide area, as shown in **Table 1**, the field survey confirmed that under-floor inundation and above-floor inundation are expected in Iwaki City. Furthermore, **Table 1** shows that the land use in Funaoka District of Shibata Town also has a high percentage of residential land. **Figure 3** shows that the area from 0 m to 5.0 m is widespread, and as shown in **Figure 4**, the depth of inundation was 1.5 m at the deepest point, confirming that most of the damage was caused by under-floor inundation.

4.2 Questionnaire surveys with local residents

4.2.1 Summary of questionnaire surveys

In this study, questionnaire surveys were conducted with residents living in the study area to understand the evacuation behavior and disaster prevention awareness of residents in the event of river flooding. **Table 2** provides a summary of the survey.

Survey Area	TairaShimohirakubo District Iwaki City	Funaoka District, Shibata Town
Date of survey	December 14, 2019 and January 5 and 13, 2020	December 15–18, 2020
Number of samples	30	45
Survey item	Respond attributes (age group, years of residence)	
	Whether or not hazard maps are checked in advance	
	Find out about designated shelters in advance	
	Potential flooding of your home as of prior to Typhoon No. 19	
	Flooding damage to home caused by Typhoon No. 19	
	Evacuation or not during Typhoon No. 19	
	Triggers and reasons regarding the presence or absence of evacuation	

Table 2.
Summary of the questionnaire survey.

	TairaShimohirakubo District, Iwaki City		Funaoka District, Shibata Town	
Age groups	Nimber of Respondents	Shares	Nimber of Respondents	Shares
10s	0	0.0%	5	11.1%
20s	0	0.0%	20	44.4%
30s	0	0.0%	2	4.4%
40s	2	6.7%	2	4.4%
50s	6	20.0%	1	2.2%
60s	10	33.3%	4	8.9%
70s	9	30.0%	8	17.8%
80s or older	3	10.0%	3	6.7%
Tatal	30	100.0%	45	100.0%

Table 3.
Age groups of respondents.

4.2.2 Results and discussion of the questionnaire surveys

Table 3 shows the age of the respondents, and **Table 4** shows the years of residence of the respondents. **Table 5** shows whether or not the respondent was aware of or had confirmed the designated shelter in their area. **Table 3** shows an age bias between respondents from Iwaki City and Shibata Town; similarly, **Table 3** shows a difference in attributes between the two regions, with Shibata Town having many respondents in their teens. However, **Table 5** shows that more than 60% of the respondents in both regions were aware of and had confirmed the existence of their designated shelter, a commonality found in both regions.

Table 6 shows how likely the respondents thought their homes may be affected by Typhoon Hagibis in 2019. **Table 7** shows the actual damages caused by the typhoon. **Table 7** shows that more than 90% of the respondents' homes were inundated above floor level, similar to the overall situation in Iwaki City shown in **Table 1**. Therefore, as shown in **Figure 2**, it is likely that kerosene heaters and futons were removed from the home during disaster cleanup in Iwaki City.

However, it was found that more than 90% of the respondents in the Tairashimo-hirakubo District thought that their area would never be inundated. In the process of conducting the questionnaire survey, when asked about the basis for ensuring their area would not be inundated, all 21 responded, "Because it never happened before" or

	TairaShimohirakubo District, Iwaki City		Funaoka District, Shibata Town	
Years of residence	**Nimber of Respondents**	**Shares**	**Nimber of Respondents**	**Shares**
1–5	1	3.3%	17	37.8%
6–10	1	3.3%	2	4.4%
11–20	7	23.3%	4	8.9%
21–30	5	16.7%	6	13.3%
31–40	6	20.0%	6	13.3%
41–50	3	10.0%	3	6.7%
51 and more	7	23.3%	7	15.6%
Total	30	100.0%	45	100.0%

Table 4.
Years of residence.

	TairaShimohirakubo District, Iwaki City		Funaoka District, Shibata Town	
Recognition of designated shelters	**Nimber of Respondents**	**Shares**	**Nimber of Respondents**	**Shares**
I know it	20	66.7%	27	60.0%
I do not know	10	33.3%	18	40.0%
Total	30	100.0%	45	100.0%

Table 5.
Advance knowledge regarding designated shelters.

	TairaShimohirakubo District, Iwaki City		Funaoka District, Shibata Town	
Possibility of suffering from the disaster ※Including flooding under the floor	**Nimber of Respondents**	**Shares**	**Nimber of Respondents**	**Shares**
I thought for sure it would flood.	1	3.3%	1	2.2%
Thought it might flood.	0	0.0%	14	31.1%
Neither	0	0.0%	8	17.8%
I thought it would not get to the point of flooding.	8	26.7%	15	33.3%
I thought it would never flood.	21	70.0%	7	15.6%
Total	30	100.0%	45	100.0%

Table 6.
How likely respondents thought their home would be affected by Typhoon No. 19 in 2019.

	TairaShimohirakubo District, Iwaki City		Funaoka District, Shibata Town	
Actual Disaster Situation	**Nimber of Respondents**	**Shares**	**Nimber of Respondents**	**Shares**
inundation above floor level	28	93.3%	5	11.1%
inundation under floor level	2	6.7%	6	13.3%
No flooding damage	0	0.0%	34	75.6%
Total	30	100.0%	45	100.0%

Table 7.
Actual disaster results from Typhoon No. 19 in 2019.

"Because the river is far away from my house." In addition, **Table 8** shows the evacuation status of the respondents.

Table 8 shows that about 70% of the respondents, especially in the Tairashimohirakubo District of Iwaki City, did not intend to evacuate or could not do so even if they wanted to. The 13 respondents in **Table 8** who did not intend to evacuate and the ten respondents who were unable to evacuate were asked about their awareness and situation during the typhoon as a free-text entry. **Table 9** describes the attitudes and circumstances of respondents who did not intend to evacuate. **Table 10** describes the attitudes and circumstances of the ten persons who could not evacuate.

	TairaShimohirakubo District, Iwaki City		Funaoka District, Shibata Town	
evacuation situation	**Nimber of Respondents**	**Shares**	**Nimber of Respondents**	**Shares**
Evacuated	7	23.3%	9	20.0%
I did not evacuate.	13	43.3%	29	64.4%
I was unable to evacuate.	10	33.3%	7	15.6%
Total	30	100.0%	45	100.0%

Table 8.
Evacuation status during Typhoon No. 19 in 2019.

Respondents' attitudes and behavior	Nimber of Respondents
I thought I would be safe if they evacuated upstairs.	6
There was no certainty that water would come.	1
I felt safer at home than during the evacuation on the road.	1
As I had never experienced anything like this before, I thought there was no need to evacuate.	1
I thought it was okay because it's surrounded by rice fields and the river is far away.	1
I have a dog and I was at home to protect it.	1
I had a cat and was at home to protect it.	1
I simply thought it would be fine.	1
Total	13

Table 9.
Attitudes and circumstances of respondents who did not intend to evacuate.

Attitudes and circumstances of respondents who could not evacuate.	Nimber of Respondents
By the time I thought to evacuate, it was too late, so I evacuated upstairs.	8
I have a disabled family member.	2
Total	10

Table 10.
Attitudes and circumstances of respondents who could not evacuate.

Typhoon Hagibis 2019 in the Tairashimohirakubo District of Iwaki City only caused inundation depths of 3 m or less above floor level. As shown in **Table 9**, the residents escaped the flooding by evacuating vertically. However, according to hazard maps, there was a good chance that the second floor of their buildings could have been submerged. A decision based on stereotypes and wishful thinking not to evacuate would have been extremely dangerous.

Furthermore, although human life takes precedence over pets in an emergency, there is a growing debate regarding the evacuation of pets in accordance with the Act on Welfare and Management of Animals [16]. Knowledge and appropriate judgment by pet owners themselves are essential to protect the lives of pets as well. However, since evacuation instructions by local authorities must be followed, pet owners are required to consider and prepare an evacuation plan with their pets in advance.

Table 10 confirms that slow decision-making was the most common cause of residents who indicated that they intended to evacuate but could not do so. In recent years, many institutions and organizations have recommended the development of Community Disaster Management Plans and creating a "My Timeline." If you intend to evacuate, you are urged to be prepared in advance so that it is not too late. On the other hand, some residents have physical or mental illnesses that make it difficult for them to evacuate in the same manner as non-disabled persons. Together with local governments, various support groups, residents, etc., they should discuss in advance how mutual support and public support should be provided and how to respond in the event of an emergency.

5. Disaster prevention awareness of residents in sediment disaster hazard zones

5.1 Sediment disasters and the sediment disasters prevention act

Sediment disasters generally refer to "debris flow," "landslides," and "slope failures" based on the mechanism of occurrence and the movement of earth and sand. **Figure 5** shows a conceptual diagram of such sediment disasters [17]. These hazardous areas are called sediment disaster hazard zones, where hazards are publicized, and alert and evacuation systems are established to prevent and mitigate the damage caused by sediment disasters. On the other hand, these sediment disaster special hazard zones are intended to prevent facilities used by residents who require special consideration for evacuation from being located in new areas with a high risk of sediment disasters. Specifically, these hazard zones refer to zones where restrictions such as a permit system for certain development activities and structural regulations for buildings are imposed, limiting those that particularly need regulation from the development stage [18].

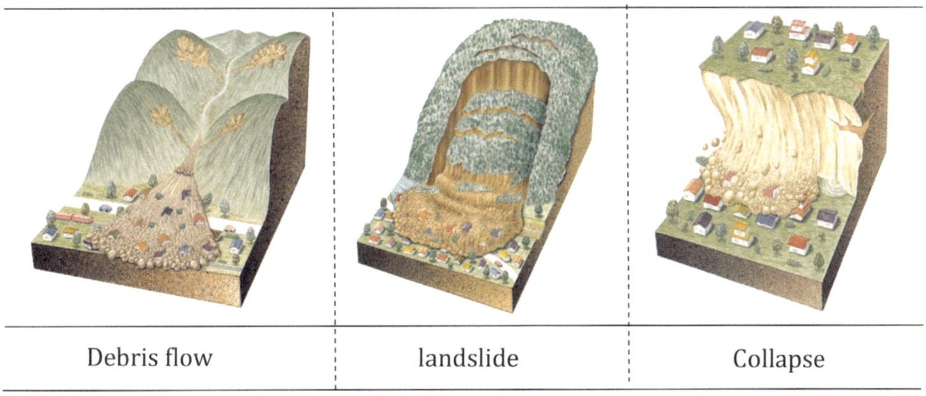

Figure 5.
Image of landslide types.

Survey Area	Matsuyama and Naruko districts, Osaki City, Miyagi Prefecture
Date of survey	28 September 2020, from 30 November to 14 December 2020.
Number of samples	8 samples
Survey item	Respond attributes (age group, years of residence)
	Find out about designated shelters in advance
	Awareness of hazard maps published by Osaki City
	Recognition of landslide hazards.
	Importance of evacuation when an evacuation order is issued.
	Willingness to change the place of residence in the future and the reasons for this.

Table 11.
Summary of the questionnaire survey.

Twenty-three years have passed since the Act on Sediment Disaster Countermeasures for Sediment Disaster Prone Areas was enacted. According to the Ministry of Land, Infrastructure, Transport and Tourism, as of the end of December 2023, there are 690,226 sediment disaster hazard zones and 591,748 sediment disaster special hazard zones across Japan [19].

Under the Act on Sediment Disaster Countermeasures for Sediment Disaster Prone Areas, prefectural governors are authorized to recommend relocating buildings that would be severely damaged in a sediment disaster in sediment disaster special hazard zones. For example, the existing system has been expanded by adding the "project to relocate houses in danger of being damaged by sediment disasters" in sediment disaster special hazard zones. This program gives grants to relocate housing in dangerous situations. However, as of 2015, 30% of nonconforming housing had been relocated, 55% had not been relocated, and 15% were unknown [20].

In this study, we conducted questionnaire surveys of residents living in sediment disaster hazard zones to understand their awareness of disaster management by interviewing them about their housing relocation.

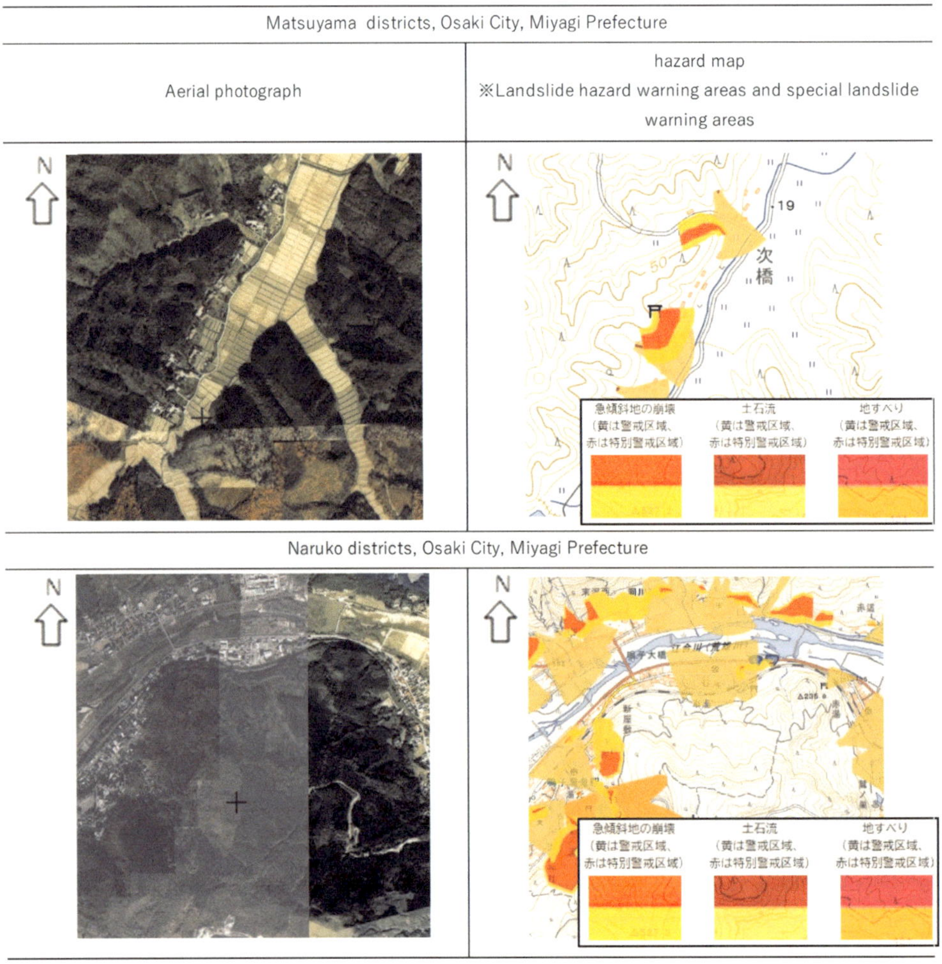

Figure 6.
Hazard map status in Matsuyama and Naruko Districts.

5.2 About the surveyed areas

5.2.1 Summary of questionnaire surveys

As noted in **Table 1**, the questionnaire surveys were conducted in the Matsuyama and Naruko Districts of Osaki City, Miyagi Prefecture. **Table 11** summarizes the study, and **Figure 6** shows the sediment disaster hazard maps for the Matsuyama and Naruko Districts.

As shown in **Table 11**, the reason why only eight samples were obtained is that there were approximately 70 residences located in the sediment disaster hazard zones and sediment disaster special hazard zones, but the questionnaire surveys were conducted during the COVID-19 Crisis. As shown in the photos in **Table 11**, in addition to wearing masks, it was necessary to keep a distance of approximately 3 m from the respondents outside their homes without entering their front doors. Therefore, when we explained the purpose of this survey and informed them of the survey method, there were many occasions when they felt that it was extremely troublesome. We assume this was why we could not obtain survey cooperation from the respondents.

5.2.2 Results and discussion of the questionnaire surveys

Table 11 shows the results of the questionnaire survey items presented in **Table 11**. **Table 12** shows the age of the respondents, and **Table 13** shows the respondents' years

	Matsuyama and Naruto areas, Osaki City.	
Age groups	**Nimber of Respondents**	**Shares**
50s	2	25.0%
60s	2	25.0%
70s	3	37.5%
80s or older	1	12.5%
Total	8	100.0%

Table 12.
Age groups of respondents.

	Matsuyama and Naruto areas, Osaki City.	
Years of residence	**Nimber of Respondents**	**Shares**
1–5	1	12.5%
6–10	3	37.5%
11–20	1	12.5%
21–30	0	0.0%
31–40	0	0.0%
41 and more	3	37.5%
Total	8	100.0%

Table 13.
Years of residence.

of residence. **Table 14** shows the respondents' awareness of flood hazard maps in Osaki City, and **Table 15** shows the respondents' awareness of designated shelters in the area. **Table 16** shows respondents' awareness that their homes may be affected by a sediment disaster, **Table 17** shows the importance of taking action when an

Matsuyama and Naruto areas, Osaki City		
Recognition of designated shelters	**Nimber of Respondents**	**Shares**
I know it	6	75.0%
I do not know	2	25.0%
Total	8	100.0%

Table 14.
Awareness of hazard maps in Osaki City.

Matsuyama and Naruto areas, Osaki City		
Check the flood hazard map	**Nimber of Respondents**	**Shares**
Have checked.	5	62.5%
Do not know, cannot remember.	2	25.0%
Never checked.	1	12.5%
Total	8	100.0%

Table 15.
Respondent confirmation of flood hazard maps.

Matsuyama and Naruto areas, Osaki City		
The presence of a home in a landslide warning area	**Nimber of Respondents**	**Shares**
I've known you for a long time.	6	75.0%
The question you just asked is the first I've heard of it.	1	12.5%
non-response	1	12.5%
Total	8	100.0%

Table 16.
Awareness of respondent's home in a landslide warning area.

Matsuyama and Naruto areas, Osaki City		
Awareness of the dangers when an evacuation order is issued.	**Nimber of Respondents**	**Shares**
Extremely dangerous.	7	87.5%
Slightly dangerous.	0	0.0%
Neither.	0	0.0%
Not very dangerous.	1	12.5%
Not dangerous at all.	0	0.0%
Total	8	100.0%

Table 17.
Awareness of the dangers when an evacuation order is issued.

Matsuyama and Naruto areas, Osaki City		
Intention to change place of residence in the future	**Nimber of Respondents**	**Shares**
Specifically planning to move.	1	12.5%
Somehow planning to move.	0	0.0%
Neither.	2	25.0%
Not much moving is planned.	2	25.0%
No plans to move at all.	3	37.5%
Total	8	100.0%

Table 18.
Willingness of the respondent to change their place of residence in the future and reasons thereto.

evacuation instruction is issued, and **Table 18** shows intentions and reasons for future change of residence.

Table 12 shows that all of the sediment disaster questionnaire survey respondents were 50 years old or older, which puts them in the elderly group. **Table 13** shows that they were polarized into two groups in terms of years of residence: those who had lived in the area for less than 10 years and those who had lived there for 40 years or more. **Table 14** shows the recognition of designated shelters in the respondents' area. **Table 15** shows whether or not the residents were aware of or have confirmed the Osaki City Flood Hazard Map. In addition, **Table 16** shows respondents' perceptions of whether or not their homes were located in a sediment disaster hazard zone.

Tables 14 and **15** show that approximately 70% of residents are aware of the existence and location of designated shelters and have checked the Osaki City Flood Hazard Map. However, several respondents indicated they did not know about checking hazard maps or designated shelters. Still, during the questionnaire survey, we learned that these respondents usually lived in a different location and happened to be respondents at the time of the survey. Therefore, it was confirmed that almost all residents in the relevant area of Osaki City, the target of this city, are aware of hazard maps and designated shelters.

Table 17 shows the importance of evacuation when an evacuation instruction is issued, and **Table 18** shows whether the respondents intend to move to safer areas in the future. Furthermore, **Table 19** describes the reasons given in **Table 18** for respondents not considering a move in the future as a free-text entry.

Table 16 shows that about 80% of the residents know their homes are in a sediment disaster hazard zone. **Table 17** shows that they are aware of the dangers when an evacuation instruction is issued. However, **Table 18** shows that more than 60% of residents do not plan to move from home to safer areas, even if they live in

I've lived here for a long time and it's where I was born.
Due to financial issues, such as the cost of construction and other activities.
Because of financial constraints.
Because I do not feel the need to.
Because my parents are close by.

Table 19.
Reasons for the respondent not to change their place of residence in the future.

such danger zones. From **Table 19**, it was confirmed that the reasons were due to the "attachment to the land where they were born and raised," "family," and "financial circumstances."

6. Conclusion

In Japan, a disaster-prone country, this study focused mainly on Typhoon Hagibis in 2019. We researched evacuation and disaster management awareness of residents due to river flooding caused by torrential rains and disaster management awareness of local residents living within a sediment disaster hazard zone. The results showed that approximately 95% of residents in the Tairashimohirakubo District of Iwaki City, Fukushima Prefecture, and 50% of residents in the Funaoka District of Shibata Town, Miyagi Prefecture, believed that their areas would not be inundated to the point of their homes getting flooded. In particular, 93.3% of Iwaki City as a whole was inundated above floor level, causing extensive damage far beyond the residents' imagination.

On the other hand, a detailed look at the evacuation status of residents reveals (1) Those who evacuated, (2) Those who did not intend to evacuate, and (3) Those who wanted to evacuate but it was already too late. Inundation in the Tairashimohirakubo District of Iwaki City due to Typhoon Hagibis in 2019 was within 3 m, and vertical evacuation was an effective option to save lives. However, the maximum expected inundation in the area is 5 m to 10 m. Recognizing the above points as one of the lessons to be learned in disaster risk management measures is essential. In addition, some residents have a physical or mental illness, among other household circumstances. This highlights the importance of consulting with various groups and designing systems for contingencies more than ever.

In addition, residents whose homes are located in sediment disaster hazard zones are aware of the danger of sediment disasters and are willing to evacuate actively. However, they were reluctant to relocate their homes in the future, confirming that attachment to the community and financial hurdles were the main factors. As we aim to create compact cities, we must urgently consider and implement sustainable adaptation measures to protect lives, including designing all policies and subsidy programs for disaster management, including those for relocation.

Acknowledgements

The laboratory members provided significant help in carrying out the fieldwork and interviews for this study. I would also like to thank the residents for cooperating with the survey. This study was supported by JSPS KAKENHI Grant Number 23 K01815, Grant-in-Aid for Scientific Research (C). I hereby note and state my gratitude to you all.

Conflict of interest

The authors declare no conflict of interest.

Author details

Teppei Ishiuchi
Department of Regional Sciences, School of Project Design, Miyagi University, Taiwa-cho, Miyagi Prefecture, Japan

*Address all correspondence to: ishiuchit@myu.ac.jp

References

[1] Japan Institute of Country-ology and Engineering. Surprisingly little is known about Japan's land [Internet]. 2023. Available from: https://www.jice.or.jp/knowledge/japan (in Japanese)

[2] Ministry of Land, Infrastructure, Transport and Tourism, 2017 Ministry of Land, Infrastructure, Transport and Tourism [Internet]. 2023. Available from: https://www.mlit.go.jp/saiyojoho/manifesto/manifesto10.html (in Japanese)

[3] Aoyama T et al. Handbook Urban Disaster Prevention, Chapter 2, Section 1. NTS Inc. 2023. p. 163 (in Japanese)

[4] Ministry of Land, Infrastructure, Transport and Tourism. White Paper on Land, Infrastructure, Transport and Tourism. Japan: White Paper on Land Transport; 2021. p. 280 (in Japanese)

[5] Fire and Disaster Management Agency (FDMA). Firefighting Movement, Major Amendments to the Basic Act on Disaster Control Measures from its Enactment to the Present, Publication April 2022. Japan: Fire and Disaster Management Agency (FDMA); 2022. p. 6 (in Japanese)

[6] Nishida G. Fire and Disaster Management Agency (FDMA), Recent Trends and Issues Concerning Disaster Countermeasure Laws-Preparing for Increasingly Frequent and Severe Disasters. In: Legislation and Research, No. 4042018. Japan: Planning and Coordination Office, House of Councilors, The National Diet of Japan. p. 101 (in Japanese)

[7] FEMA, Emergency Management in the United States. Independent Study Program IS-111, Livestock in Disasters, Unit 4 [Internet]. 2015. Available from: https://training.fema.gov/emiweb/downloads/is111_unit%204.pdf

[8] Drabek TE, FEMA, The Social Dimension of Disaster. Emergency Management Institute, Instructor Guide [Internet]. 1996. Available from: https://ntrl.ntis.gov/NTRL/dashboard/searchReults/titleDetail/PB99105488.xhtml#

[9] Nishida G. Fire and disaster management agency, recent trends and issues concerning disaster countermeasure laws-preparing for increasingly frequent and severe disasters. In: Legislation and Research, No. 404. Japan: Planning and Coordination Office, House of Councilors, The National Diet of Japan; 2018. p. 102 (in Japanese)

[10] Fire and Disaster Management Agency (FDMA). Firefighting Movement, Major Amendments to the Basic Act on Disaster Control Measures from its Enactment to the Present, Publication April 2022. Japan: Fire and Disaster Management Agency (FDMA); 2022. pp. 12-14 (in Japanese)

[11] Japan Meteorological Agency (JMA). Heavy Rain, Storm, Etc. Due to Typhoon No. 192019. Japan: Japan Meteorological Agency (JMA); p. 1 (in Japanese)

[12] Iwaki City Disaster Control Headquarters, Disaster occurred on October 12, 2019 'East Japan typhoons, etc.', Disaster Report. 2022. Available from: https://www.city.iwaki.lg.jp/www/contents/1572144227297/simple/saigaihoukoku220228.pdf (in Japanese)

[13] Shibata Town, Miyagi prefecture, Typhoon in East Japan in the first year of

the order and year (Typhoon No. 19 of 2019), overview of damage and disaster control headquarters response. 2022. Available from: https://www.town.shibata.miyagi.jp/index.cfm/75,39913,127,html

[14] Aerial photograph, Geospatial Information Authority of Japan, Geographical Survey Institute Map' Electronic National Land Web'. Available from: https://maps.gsi.go.jp/

[15] Ministry of Land, Infrastructure, Transport and Tourism of Japan. Hazard map portal site-overlapping hazard maps. Available from: https://disaportal.gsi.go.jp/

[16] Ministry of the Environment. Disaster Preparedness Guidelines for People and Pets. Japan: Ministry of the Environment; 2018. pp. 44-49 (in Japanese)

[17] NPO Sediment Disaster Prevention Publicity Center (SPC). Learning and communicating - A collection of illustrations. 2001. Available from: https://www.sabopc.or.jp/sozai/sozai_category/illust_cate/#id27 (in Japanese)

[18] Ministry of Land, Infrastructure and Transport. Outline of the landslide prevention act, The first earth and sand disaster prevention countermeasures subcommittee. 2019. p. 1 (in Japanese)

[19] Ministry of Land, Infrastructure and Transport. Erosion control department, designation of landslide hazard warning zones, etc., throughout the country, as of March 31, 2024. Available from: https://www.mlit.go.jp/mizukokudo/sabo/content/001741207.pdf (in Japanese)

[20] Ogawa K. Changes in landslide prevention legislation and the 'mission' of geospatial information technicians. The Journal of Survey. 2024;**5**:28 (in Japanese)

Chapter 3

Understanding the Evolving Nature of Urban Flood Risks in Sub-Saharan Africa: The Case of Kampala City, Uganda

Denis Arinabo

Abstract

The escalation of urban flood occurrences in Sub-Saharan African cities necessitates a nuanced comprehension of evolving flood catalysts beyond the realms of large-scale rapid urbanisation and climate change discourse. Whilst current research and policy discussions primarily focus on these aspects, urban flooding remains a persistent challenge for stakeholders. This chapter addresses this disparity by presenting a precise alternative conceptual lens for understanding changing flood triggers. Through a multi-method approach involving key informant semi-structured interviews and web-based surveys supported by policy review and GIS, the study examines Kampala City as a case study. The findings highlight the intricate interplay of socio-economic, institutional, environmental, and infrastructural factors within diverse socio-spatial contexts and urban dynamics and processes. By emphasising context-specific drivers, the chapter advocates for tailored flood risk policies, urging urban authorities to move beyond 'one-size-fits-all' approaches for sustainable flood risk management in African cities.

Keywords: changing flood drivers, context-specific factors, wetland loss, slum incursion, indiscriminate waste disposal, Kampala city

1. Introduction

Urban flooding, a growing issue of concern in many cities and urban areas, has drawn attention due to its profound effects on human health, buildings and the socio-economic fabric of communities [1]. Across various geographical contexts, urban flooding manifests through various triggers and causes, including fluvial, pluvial, groundwater, coastal and flash floods [2], which interact with both natural and human forces, resulting in widespread devastation [3, 4]. Globally, recent catastrophic urban floods in the 21st century are linked to tsunami and heavy rainfall, particularly in countries with expansive river basins that receive extreme precipitation events such as Mozambique, Brazil, India, Pakistan, Bangladesh, Niger, Ghana (Accra), Zambia, Togo and Colombia [2, 5]. Whilst most of these triggers are caused by natural events, the magnitude of the impact is aggravated by human actions. For instance, the cases of El Nino phenomena in countries like Colombia and Uganda have often coincided with extensive deforestation on mountain slopes, leading to rain-saturated land movement or mudslides [6, 7]. Consequently, such sedimentation raises water levels in the nearby water bodies such as rivers, thereby causing spontaneous flash floods in neighbouring urban areas.

In the Sub-Saharan Africa (SSA) context, floods are the most frequently occurring natural disaster [2, 8], having hit a breaking record of up to 654 flood events recorded between 1980 and 2013 [5]. This figure could have been doubled considering the number of sporadic flooding events experienced in the last decade. For example, according to the United Nations Office for the Coordination of Humanitarian Affairs (OCHA), severe flooding situations affected more than 2.2 million people in 2020 compared to 1.1 million people in 2019 in 11 West and Central African countries [9, 10]. This excludes other isolated widespread flooding events in SSA cities such as the Free-town mudslide floods of 2017, the Nairobi floods of 2019 and Kampala's worst floods between 2017 and 2019, among others.

The commonalities of the increased regional exposure to evolving urban flood risks across SSA cities have been attributed to various theories including but not limited to large-scale urbanisation trajectories [11]. In this, the increase in the occurrence of indiscriminate flooding is attributed to the increase in urban populations which triggers the urbanisation processes that directly or indirectly contribute to flooding experiences [12]. Particularly, urbanisation initiates several changes that have an overall net effect on the natural environment whilst reducing permeable surfaces for flood waters to escape. Consequently, this increases flood risk and vulnerability of urban communities in the face of usual tropical precipitation events or extreme weather conditions experienced across the region.

As it is with rapid urbanisation trajectories for African cities, the events that precipitate floods remain diverse, multifaceted and interrelated. As a result, the academic and policy discourses have a weak and insignificant cause-link relationship between flood experiences and drivers [13]. This may be attributed to the lack of flood risk benchmarks in low and middle-income countries that characterise the region. Other flood drivers are attributed to proximity to water bodies and precipitation changes in constantly changing urban spaces [2, 14, 15]. Besides, other ingredients have continually been associated with worsening flood experiences. These among others include colonial legacies, endemic poverty and lack of governance capacity, for example, financial resources, high levels of informalisation, low levels of economic development and contested urban development processes [11, 16, 17]. Nonetheless, it is a widely accepted phenomenon that flooding in many African cities such as Kampala, Nairobi, Bujumbura, Abuja and Kigali is triggered by rainfall.

Besides, previous research indicates that the largest cities particularly those along large water bodies areas are more at risk of fluvial flooding as a result of sea rise [18]. In the East African region, however, there is still a disparity regarding what causes widespread flooding occurrences in cities along the shores of Lake Victoria, the largest inland freshwater lake in Africa. The Inter-Governmental Panel on Climate Change (IPCC) projected that the East African region would receive extreme precipitation resulting in sea rise, yet what is known is that the coastline (in Tanzania) and the water levels in Lake Victoria have been falling [5, 19] up until recently when the lake experienced an all-time high water level rise between 2019 and 2020 since the highest record of 1964. Whilst there is still limited empirical research regarding the cause of water level rise, it is argued that poorly managed urbanisation, heavy precipitation and land use change within the lake basin are responsible for the increase in water levels in Lake Victoria [20, 21].

Beyond pluvial flooding, SSA shares a long history of floods particularly fluvial floods associated more closely with changes in river discharges downstream, but all the same, triggered by heavy rainfall in the upper catchment. Rivers such as the Nile floods large parts of Egypt in Northern Africa; similarly, rivers such as Zambezi, Niger, Congo and Limpopo cause floods in their respective geographical locations. Across the globe, fluvial flooding situations have been felt in many areas, for example, the outburst of the Yangtze River in China and Mississippi in the United States [2]. Because of the long history that relates human settlement to water sources such as rivers and lakes, most urban centres and cities in Africa are concentrated along these features, increasing their flood risk vulnerability to fluvial floods. In West Africa, such floods have been commonly experienced in cities within the Niger Delta [14] and the Congo Basin [15]. Such flood events, irrespective of the cause, continue to pose recurrent threats to both human populations and the urban environment as well as additional strain on the already fragile urban economies in SSA [11].

Managing current and future urban flood risks becomes more complex in SSA where rapid urbanisation is accompanied by a proliferation of informal settlements. These subject more assets, teleconnections and urban infrastructure to urban flood risks [12]. Such urban dynamics have made it difficult to precisely quantify historical flood impact in many African cities. Worse still, in most cases, local knowledge and community perceptions have been disregarded by urban governments due to prejudices [22]. However, whilst it may be easier to apportion blame on urban government and institutions, research shows that people also directly contribute to the flooding problem in various ways such as through indiscriminate waste disposal and floodplain occupation [12, 23].

Inevitably, flooding remains a multiplex issue, and its impacts are wide-ranging across African cities. The uniqueness of the African urban space, therefore, requires unveiling the changing context-specific urban conditions that exacerbate flood risks. This is largely missing across much-published research on changing flood risk drivers in SSA. This chapter therefore advances a precise and holistic understanding of the complexities surrounding urban flooding in SSA cities, using Kampala as a case study. The chapter takes into consideration the historical legacies premised on colonial urban planning and development practices and further interrogates the prevailing contextual urban socio-spatial conditions, that is, informal settlements, loss of wetlands and indiscriminate poor waste disposal, that characterise flood hotspots.

2. Research setting and methods

Uganda's capital Kampala, one of the largest cities in the East African region, alongside Nairobi (Kenya), Addis Ababa (Ethiopia) and Dar-es-Salaam (Tanzania) was selected as the case study following the worst flood events experienced between 2019 and 2020. With a population growth rate of over 5%, which is above the African average of 4.71% (see **Table 1**) and way above the global average of 2.5% [3], Kampala remains among the top 10 rapidly growing cities in SSA whose population is expected to double by 2030 [24]. The city has also undergone unprecedented large-scale urban expansion due to rapid demographic changes in the last 2 decades. This rapid expansion has been accompanied by a proliferation of highly congested slum settlements [25] that double as flood hotspots [12]. This trend is common among many SSA cities (see e.g., [26–28]), with the informal settlements forming the bulk of the semi-urbanised areas that house the majority of the urban population. In Kampala, slums house approximately 560,000 families across 57 informal settlements that

Country	Urban agglomeration	2000 population (thousands)	2020 population (thousands)	Annual average population growth rate (2000–2020)	Annual increment in population 2000–2020 (thousands)
Nigeria	Abuja	833	3,278	7.1%	122
Burkina Faso	Ouagadougou	921	2,780	5.7%	93
United Republic of Tanzania	Dar es Salaam	2,272	6,702	5.6%	221
Cameroon	Yaoundé	1,351	3,992	5.6%	132
Angola	Luanda	2,829	8,330	5.5%	275
Ghana	Kumasi	1,187	3,348	5.3%	108
Nigeria	Port Harcourt	1,091	3,020	5.2%	96
Uganda	Kampala	1,233	3,298	5.0%	103
Cameroon	Douala	1,490	3,663	4.6%	109
Madagascar	Antananarivo	1,361	3,369	4.6%	100
DR Congo	Kinshasa	6,140	14,342	4.3%	410
Kenya	Nairobi	2,214	4,735	3.9%	126
Ethiopia	Addis Ababa	2,377	4,794	3.6%	121
Nigeria	Lagos	7,281	14,368	3.5%	354
South Africa	Johannesburg	3,046	5,783	3.3%	137
Côte d'Ivoire	Abidjan	3,007	5,203	2.8%	110
Sudan	Khartoum	3,505	5,829	2.6%	116
South Africa	Cape Town	2,785	4,618	2.6%	92
Egypt	Cairo	13,626	20,901	2.2%	364
Egypt	Alexandria	3,546	5,281	2.0%	87

Source: [24], 2022.

Table 1.
Population size and growth rates for Kampala and other African cities between 2000 and 2020.

spread to the broad valleys originally covered by wetlands [12, 25]. In Dar es Salaam where urbanisation is also manifesting informally, slums house more than 70% of the urban population [29]. Such rapid slum formations coupled with years of neglect and underinvestment in major infrastructural projects and municipal services have led to drainage infrastructural deficits, further presenting a unique set of challenges to flood management in the SSA region.

The selection of Kampala city as the case study was anchored on its contextually challenging urban dynamics that affect the nature of urban flooding. Across Kampala's flood hotspots (see **Figure 1**), recurrent and widespread urban flood risk is principally caused by extreme rainfall events spread out over two rainy seasons, that is, from March to May and from October to December, with an annual average of 1406 mm. Previous research indicates that whereas Kampala's rainfall variability and weather changes are primarily influenced by Lake Victoria, there has been a general decrease in the amount of rainfall received over the last decades, that is, 12% since 1983 [30]. Meanwhile, it is indicated that Kampala's flooding ordeals started in the early 1990s, and the number of convective storm rainfalls responsible for urban flooding has since increased from five in 1993 to ten in 2014 [12]. Based on the general decrease in precipitation, it is clear that what exacerbates the severity and extent of floods remains a function of multiple urban conditions such as demographic changes, rapid urbanisation, informal settlements and encroachment on ecologically sensitive flood plains. This subjects more than half a million people to recurring flood risks across the hotspots, as shown in **Figure 1**. It should be noted that the emergence of the hotspots across the city is not generic, but it is rather linked with key urban processes and trends. These are further discussed in Section 3.

In terms of governance and institutional arrangements, the historical urban planning legislation introduced in 1912 by the British has continued to shape a bifurcated land regulation system and condition urban land use [17]. However, it is imperative to note that the city has undergone rapid post-independence governance reforms, that is, from the traditional systems to current central government management. The latter led to the creation of the Kampala City Council Authority (KCCA), a government-funded agency charged with overseeing the planning and development of the city. This is thought to have led to the politicisation of Kampala's urban setting, giving an edge to the ruling party over the opposition, and ultimately subverting broader systems of city governance.

To mitigate against flood risks, the government has so far taken various measures such as investing in drainage infrastructure improvement, waste management and wetland protection, among others, although their efficiency is still questionable. Recently, KCCA launched the drainage master plan (DMP), a flood management regime to tackle the widespread and recurrent flood risks across the five administrative boroughs in 2015. The 40-year DMP is intended to avert the environmental challenges associated with stormwater runoff primarily through wetland preservation

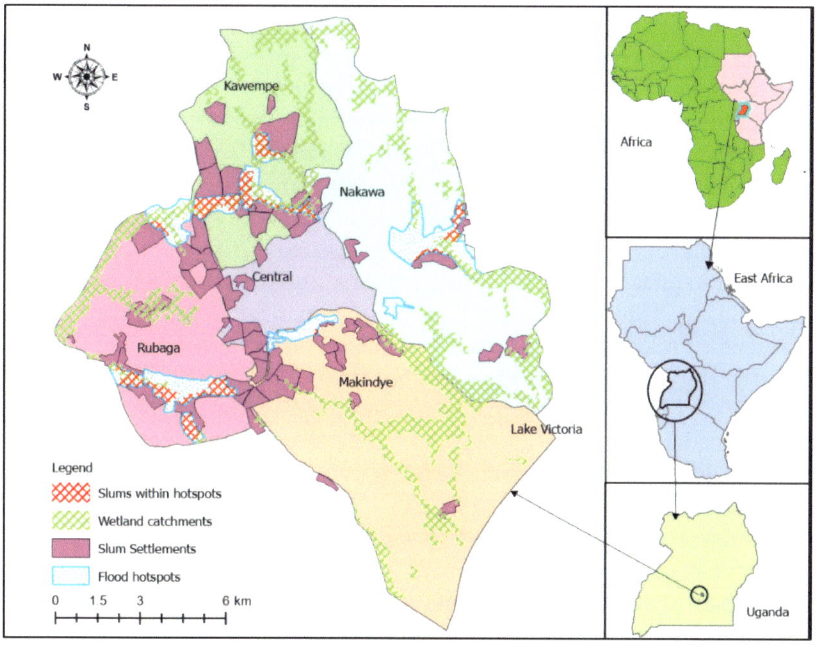

Figure 1.
Contextualising the location of Kampala City, Uganda, the extent of flood hotspots and their environs (Source: Author, 2024).

and primary and secondary drainage channel widening [12, 31]. However, with the worst flood events recorded between 2019 and 2020, a lot is still desired concerning the effectiveness of the DMP's first phase of operation (2015–2019) as intense thunderstorms render the primary and secondary drainage networks insufficient, thereby causing recurrent flash floods.

2.1 Methodology

Investigating the rampant nature of flood risks requires a precise understanding of the multiplicity of contextual urban conditions and processes that characterise SSA cities. In line with this, multiple methods, that is, detailed web survey, semi-structured key informant interviews, GIS and document review, were adopted to empirically investigate the complexities behind the evolving nature of urban flood risks across Kampala's hotspots. To achieve this, the methodology was divided into two phases.

In phase one, document or policy review in which legal, institutional, programmatic and fiscal documents closely linked with Kampala's flood risk management were reviewed. This was purposed to (a) contextualise the legal and institutional frameworks within which flooding is understood and managed and (b) generate appropriate questions for the web survey and key informant interviews. On the other hand, GIS was used to generate primary data on flood hotspots and their surroundings based on KCCA's state of urban flooding report published in 2019 as well as temporal patterns of key urban processes using Landsat remote sensing imagery for three-time series data points of 1995, 2010 and 2019. Specifically, Landsat provided insights into built-up land, vegetation and water aspects of the case study area between 1995 and 2019. To improve the accuracy of Landsat imageries and further investigate context-specific conditions within flood hotspots, secondary data sources solicited from Open Street Maps (OSM), KCCA GIS department and ACTogether Uganda were used. Preliminary findings from GIS maps played a key role in defining survey questions surrounding key features observed within and around flood hotspots.

Phase two involved primary data collection using a detailed qualitative web-based survey and key informant semi-structured interviews with participants recruited across varied institutional levels. Statistically, data were solicited from 69 purposively sampled participants, 54 of whom participated in the survey and 15 in key informant interviews. The list of these participants was generated during the preliminary policy review exercise in phase one, considering persons' and organisation's expertise, knowledge and experience regarding flooding risk management in Kampala. For local representation, community leadership in areas affected by floods was contacted. During the survey, the participants were asked to state to which category they belonged. Out of this, eight participant categories were recorded, that is, government/civil servant, academia, consultancy, NGO/CBOs, development partners, community representative, independent researcher and private sector. Based on the experiences and expertise of individual participants, data on the changing nature of floods in terms of severity, frequency and extent as well as institutional involvement in flood management were captured.

Upon realisation that not all questions were suitable for the web-based survey, the key informant semi-structured interview method was designed as a follow-up on some of the key issues not covered in the previous methods. This was intended to facilitate an in-depth investigation of context-specific flood drivers within Kampala's flood hotspots as presented in the next section.

3. Findings and discussions

3.1 Categorisation of identified flood drivers

One of the objectives of this research was to propose a precise understanding of the flood drivers in SSA cities using Kampala's case study. In line with this, web survey participants were asked to list multiple causes of urban flooding in Kampala's hotspots based on their knowledge and experience without necessarily placing them in order of significance. A myriad of issues were raised concerning urban flood causation in Kampala (see **Figure 2**). The rampant loss of wetlands, increase in the built environment, poor waste disposal, housing demand and institutional weaknesses (e.g., weak enforcement, monitoring and implementation of projects) across the city scale emerged among the most mentioned flood drivers. On the other hand, lack of research, poverty and limited awareness of waste management featured among the least mentioned drivers. To some extent, this finding corroborates those of previous research across SSA cities where most of the lumped environmental factors such as loss of wetlands, rainfall pattern changes and fast-paced transforming urban environments have been broadly studied under recent trends in urbanisation and climate change praxis [2, 13, 18].

The cataloguing of the flood causes in Kampala followed a grouping exercise that collated codes of similar relevance into a single theme. Out of the responses solicited, the collated data were grouped into four themes, that is, environmental, socio-economic, institutional, and infrastructural drivers, as shown in **Figure 2**. For purposes

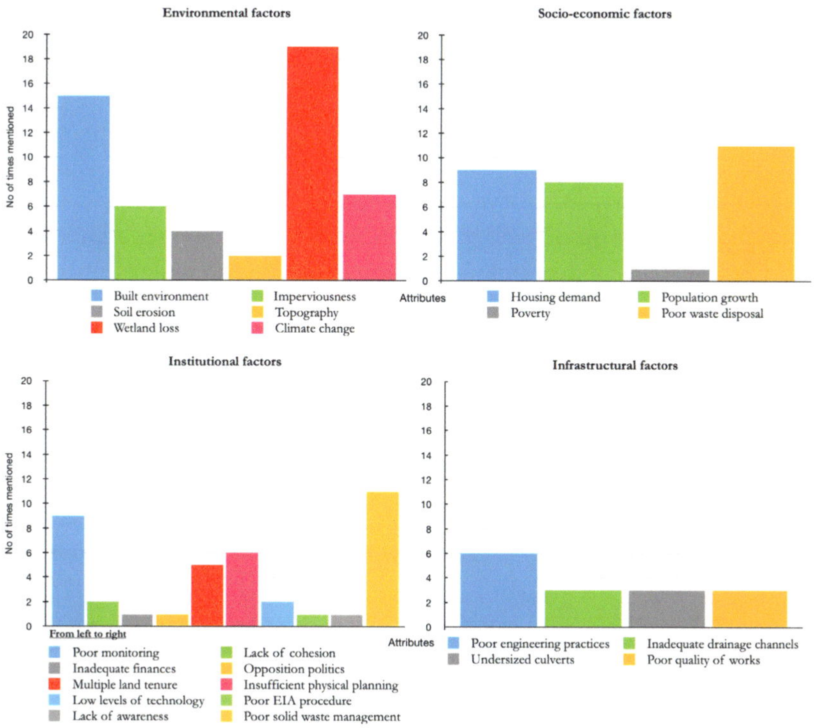

Figure 2.
The categorisation of flood drivers into themes based on the number of times mentioned (Source: Author, based on web survey data).

of this research, it should be noted that poor solid waste management as a flood driver was categorised as both an institutional and socio-economic issue mainly due to inadequate and poor management of waste collection facilities in Kampala, which is an institutional responsibility of KCCA.

Across African cities, the urbanisation trajectories have orchestrated financial constraints amidst contradictory political organisations, poverty, lack of governance capacity, high levels of formalisation, contested urban development processes and a lack of support for citizens [11, 16]. This is also seen in Kampala's case where it is believed across urban actors that rapid spatial expansion has in many ways instigated ugly scenarios that culminate into worse flood experiences. For instance, residents' rampant illegal dumping of solid waste in prohibited areas such as drainage channels, particularly those in informal settlements who lack the ability and willingness to pay for private-owned waste collection services, is linked to the government's inability to enforce planned development across city scales.

To better understand the leading causes of flooding in Kampala and to explore the interconnections between flood drivers, the individual attribute scores of every single driver within the above themes were aggregated and later expressed as a percentage of the total attribute score. The findings indicate that the environmental factors were felt to be the leading causes of urban flooding, with a 43.4% contribution, whilst socio-economic aspects accounted for the least (13.3%), as shown in **Figure 3**. Among the environmental concerns, the loss of wetlands is consistently mentioned. This reality of decreasing wetland and natural urban vegetation accompanied by sporadic flood risks is shared by many African nations, for example, Tanzania, Ghana and Nigeria [26, 29].

Beyond individual citations of flood drivers, participants were asked to assign a value that captures the significance of a particular generic flood driver as presented in a Likert scale within the survey. This was intended to draw a deeper understanding of the changing flood drivers. During the analysis, data visualisation and a ranking function in Excel were used to determine the frequency of every flood driver based on participants' choices of values assigned to each attribute from 'Not significant' represented by 1 to 'Very significant' represented by 5. The higher the value, the more relevant a flood driver is perceived. As shown in **Figure 4**, the findings indicate that wetland loss, poor waste management and inadequate drainage infrastructure are the topmost influencers of flooding in Kampala; meanwhile, inadequate funding, lack of

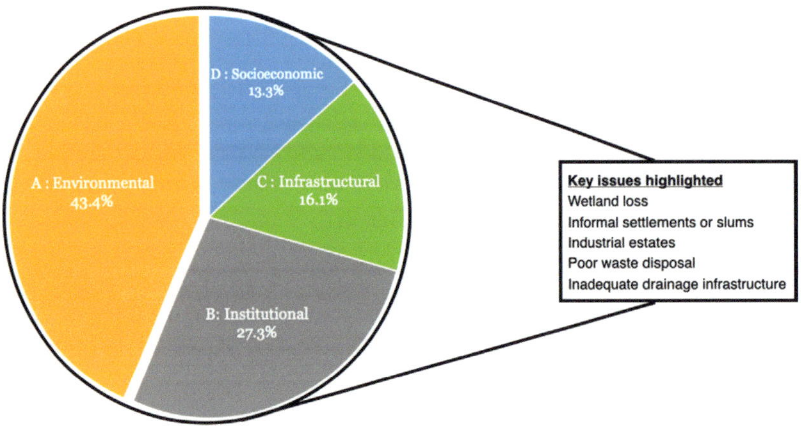

Figure 3.
Perceived categorisation of key flood drivers (source: Author, 2022).

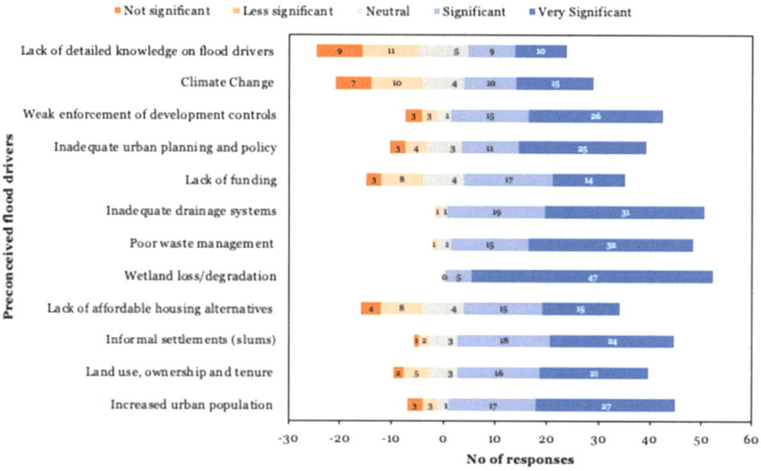

Figure 4.
Likert scale data on flood drivers for Kampala (source: Author, 2023).

affordable housing and climate change emerged as the least influential contributors to urban flooding as perceived by the survey respondents.

Triangulating findings from all data sources as presented above revealed that the loss of wetlands maintains its centrality as the topmost flood driver in Kampala. Other key contributors to urban flooding identified are associated with the increase in the built environment particularly the informal settlement incursion. It is out of this increase in the built urban environment that issues of poor waste disposal and inadequate drainage infrastructure arise. These are further amplified in informal settlements where the urban poor live. Building upon the findings of this research, the broader categorisations of flood drivers behind the evolving nature of flood risks in Kampala are further expounded in the next section.

3.2 Environmental flood drivers

Among environmental flood drivers, wetland loss emerged as the most frequently mentioned attribute. The reasons behind encroachment onto wetlands are multi-dimensional. Firstly, with an increase in urban population and without a clear physical development plan, ecologically sensitive areas such as wetlands became a target for unplanned settlement incursion. Recent research indicates that the informal settlements established in wetlands and former flood plains constitute a quarter of the city's total area, housing over 60% of the urban population [25]. This population mainly comprises the urban poor and new rural-urban migrants who target the wetlands. In other cities, the influx of people originates from people moving from other urban centres into the city and a high birth rate among the inhabitants (see e.g., [3, 29]).

Worse still, economic activities such as industrial estates, urban farming, car washing bays and brick laying have over the years also encroached on sensitive ecological environments such as wetlands, attracting poorly paid manual labourers to the areas close to these areas. With time, due to the lack of housing facilities and weak urban planning responses, the population resorted to establishing temporary haphazard settlements within the riparian areas for easy access to water. At some point, malaria became rampant in Kampala due to water-logged conditions that acted as conducive breeding environments for mosquitoes within these settlements. Consequently, the draining and conversion of wetlands were encouraged to promote malaria epidemic control [38]. Additionally, the ambiguous wetland boundaries, the plurality of land tenure and ownership systems and the failure of government regimes to regulate and implement development controls encouraged an informal housing boom in wetlands. Concerning this, a participant recounted:

> *"Few parts of Kampala such Bwaise and Nabweru were flooding in 2001 until people started settling in wetlands for example Lubigi. To date after every rainy day, flood aftermath follows in almost all former wetland areas of Kampala" (Civil Servant 7, Web-based Survey).*

This seeming policy failure could be attributed to various reasons. It is argued that the government uses confrontational, inconsistent and non-participatory government approaches to the problem of wetland encroachment characterised by forced evictions, demolitions and relocation [12, 31]. Similar cases of weak municipal interventions have been reported in other cities [3]. For Kampala, weak approaches are compounded by uncoordinated instructions between government agencies, for example, KCCA and other government institutions (such as National Environment Management Authority—NEMA and wetlands department), alleged corruption, political interference and conflicting land use policies [12].

This multiplex nature of the issues has allowed the indiscriminate conversion of wetlands, greenbelts and flood plains to go on unabated for a long time in African cities. At the city scale of Kampala, this has led to a 49.1% reduction in vegetation cover in the last 25 years as observed through Landsat imagery data point series of 1995 and 2020 (see **Figure 5**). The encroachment is primarily induced by the increase in the built environment. In this case, in the same period, that is, between 1995 and 2019, Kampala's built environment nearly doubled from 40.7 to 78.3%, respectively.

It is worth noting that the indiscriminate access to ecologically sensitive areas is not about the absence of law but rather its suspension in terms of ineffective regulation, monitoring and enforcement. In many cases, urban authorities have been found guilty of violating their own laws and not being able to enforce land use and ecosystem protection laws. For instance, in Kampala, there are sufficient laws on wetland

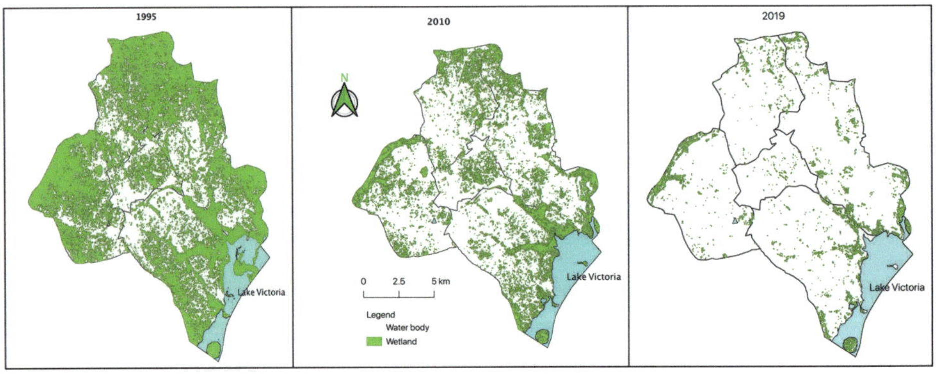

Figure 5.
Kampala's changes in vegetation cover in the last 25 years (source: Author, 2023).

protection and water catchment zones under the Ministry of Water and Environment (MWE), which also apply to Kampala city, but enforcement is severely lacking. In some cases, the practice of alienating urban land (including wetlands) to businessmen and investors is still upheld not only in Kampala but also across postcolonial Africa [32]. Such loopholes in the law have been exploited to perpetuate indiscriminate access to wetlands through alleged prevalent corruption tendencies even though stricter guidelines and penalties are in place. This manifests in the sense that whilst wetlands and their buffer zones have been demarcated, the conversion of ecologically sensitive areas still goes on unchecked.

Preserving the wetland systems is seen as a government responsibility rather than an individual effort not only in Kampala but across Uganda. Individuals across the participant categories are aware of the need to protect wetlands as buffer zones among other mitigation measures if urban flooding is to be properly managed. To this, the phrase "remove encroachers" is littered across web survey responses. However, with the increasing number of people already settled in former wetlands and flood plains, the practicality of resettling such numbers remains a protracted struggle. Unfortunately, the commonest way to remove people from flood-prone environments is via forced eviction, which is usually characterised by demolition exercises, brutality, confrontation and fierce resistance [3, 12].

Regarding climate change as an environmental flood driver, participants believed in its insignificance in influencing flood risks in Kampala due to limited evidence that supports narratives of the overall increase in the amount of precipitation received across Kampala. The Climate Change Department under MWE indicated that there were no robust projections regarding the frequency and intensity of heavy rains, even though previous research has suggested that these have been increasing and will increase with climate change [12]. Most of the studies that causally associate climate change with urban flooding would have been more useful if they demonstrated changes in rainfall measurements over a long period. When asked about climate change and urban flooding in Kampala, a consultant who has lived in Kampala city since the early 1990s remarked,

> *"I have lived in Kampala for well over 30 years and in my opinion as a Civil Engineer, urban flooding in Kampala is not about the weather changes but as a result of the reactionary approach to floods at the far end rather than at the beginning."*
> *[Consultant 6, Key Informant Interviews]*

3.3 Institutional weakness

The institutional flood drivers revolve around poor implementation of regulations by the key government institutions that are charged with managing city land use, physical planning and development. In particular, the issues raised under the institutional theme category as earlier shown in **Figure 2** include poor implementation of policy for example, managing multiple land tenure systems, poor monitoring, financing and a lack of (physical) urban planning and development. These have acted as the springboard for other flood causes, particularly the environmental and infrastructural flood drivers. Recent research has questioned the effectiveness of government institutions' involvement in flood management in SSA cities [22]. It is within this context that the role of urban institutions has been defined by a lack of clarity and disharmony in which they tend to define flooding problems in their terms, hence providing what is possible rather than what is needed [3, 22].

That said, it is widely accepted that the main challenge limiting institutional efforts is the inadequacy of financial resources required to provide and sustain key institutional operations in flood management. This finding is corroborated at the regional scale of African cities where it is acknowledged that urban authorities are confronted with increasing demands in flood risk management amidst limited logistics [3, 12]. To this, one of the key informant interviewees commented that,

> *"Planning and execution of effective urban drainage infrastructure is affected by the resource envelope which has always been the case with central funding kind of management arrangements." [Civil Servant 3, Key Informant Interviews]*

At the city level of Kampala, the above opinion was shared by many respondents, particularly consultants who have worked with KCCA on urban flood management-related projects and a few civil servants. The latter re-echoed that the government alone lacked enough budget to support most of the proffered flood management recommendations. On the contrary, however, recent findings indicate that urban institutions have in the last decade gained access to improved budgets as seen in the case of Kampala [22]. The contrast of opinion emanates from a lack of reconciled institutional responses and multiple forms of flood risk knowledge that inform flood risk management regimes.

As to why flooding is still wreaking havoc in many urban communities, the participant category of consultants was concerned about the lack of continuity of flood management projects jointly funded by KCCA and its partners such as the World Bank. Such projects usually end due to lack of funding and not necessarily upon completion. During the interviews, a consultant noted that:

> *"Once the project is finished, everything is closed, project cars and computers are sold away, and employee contracts terminated. The next project comes with new staff and starts from scratch. Funding is therefore such a big factor. Lack of continuity of funding affects many things including lesson learning." [Consultant 6, Key Informant Interviews]*

However, it should be noted that beyond government civil servants and consultants, participants in other categories expressed concerns relating to political interference, the need for a better quality of works on drainage channels and a comprehensive physical plan. It was also noted that there was a lack of cohesion among various actors, that is, development partners, government agencies and affected communities. This organisational fragmentation suggests that the financial resources available could have been enough to lessen flood consequences had there been fewer of the above-mentioned concerns and more action points.

3.4 Infrastructural crisis

The infrastructural challenges affecting flood management regimes in SSA cities can be categorised into two subsections: poor engineering practices and the quality of works when implementing drainage and related infrastructure. These have been cited concerning Kampala's ongoing DMP, a 40-year flood management regime purposed to lessen flood impact within the city. The DMP is based on 'best engineering practices', with a primary focus on widening primary and secondary channels and protecting wetlands [12, 22]. It was, however, thought by some interviewees that the DMP's main activities were 'rather unfortunate and sad' and reflected poor engineering practices since they increased the velocity of runoff and amplified its destructive power downstream (Consultant 6, Personal Interviews). Instead, they argued, the approach should be more about reducing the velocity of water by stepping the channels and making them shallower.

The impetus for implementing the DMP is the urgency to drain runoff water as quickly as possible into Lake Victoria. The policy review findings further indicate that there are plans as part of the wider DMP to contain stormwater by building check dams to ensure the slow release into receptors. However, this research could not substantiate this as there is no mention of constructed dams as part of the DMP first-phase implementation. Meanwhile, whereas the widening of primary channels is acknowledged across the data, there has been no significant improvement in flood experiences, reportedly due to constrictions within the drainage system such as culverts and lack of space to widen secondary and tertiary channels upstream. Widening primary drainage channels is further challenged by the scale of (informal) settlements and industrial estates across Kampala's flood hotspots. To the former, it has been argued that the nature of communities (in this case informal settlements) attracts a specific type of flood management regime [3]. However, the scale of informality in many SSA cities presents a multifaceted challenge that requires a precise understanding of the key issues per time in a given community.

The survey results for Kampala further indicate that structural measures such as storage reservoirs, flood embankments, secondary and tertiary drainage channels, detention basins and anti-erosion works have not fully accompanied the drainage masterplan upgrade. Even though the wider benefits of sustainable drainage systems (SUDS) in SSA have recently been examined [33, 34], some web survey respondents argue that SUDS in Kampala have not been given attention as part of flood management measures. Instead, the ongoing DMP implementation is anchored on de-siltation as one of the main activities that would help improve the state of the drainage channels [12]. However, there is a tendency to neglect the safe disposal of silt removed from the channels; it is rather heaped on the sides of the drainage channels only to be washed back into the channels when it rains. This was raised as a concern in the web-based survey by a participant who categorically wrote,

> *"Drainage channels should be occasionally opened up and waste, stones and soil from there should be dumped in safe places not just heaping it on the sides of the drainage channels because when it rains the heaped waste finds itself again in the drainage channel making it to get filled so fast and this causes an overflow of water thus floods." [Community Member 3, Web-based Survey]*

The ongoing DMP does not mention much about non-structural measures such as flood forecasting, flood plain zoning, flood proofing and disaster preparedness. For instance, when asked about the existence of flood maps for Kampala, the majority of the web survey respondents acknowledged a lack of awareness about the existence of such maps. However, a few participants who are aware of the maps mentioned that the maps are (a) only available at the City Hall at KCCA where students and planners can access them, (b) available when relevant offices are approached and (c) occasionally available in volumes of Environmental and Social Impact Statements (ESIA). Furthermore, the findings from the semi-structured key informant interviews indicate that flood risk maps are available for only a few places, that is, the Gaba-Katogo wetland catchment and Bwaise II areas. Based on the web survey participants' views, it is widely accepted that the flood maps are not widely published nor are they positioned in key places. One of the respondents commented that:

"The lack of availability of such information to the grass-root leaders (LC 1) in flood hotspot areas might also be another source of the problem" [Community Representative 4, Web-based survey].

As established above, a lack of awareness and access to flood maps denies different actors an opportunity to contribute to better flood management policies. As has been argued elsewhere, maps, when positioned openly, can be used as potential instruments for engaging the community in more creative policymaking [35]. However, for Kampala, the flood maps remain accessible to very few people which raises conceptual questions regarding their role in the overall flood management strategy. When asked during the key informant interviews, one of the consultants in response to flood maps availability argued that flood maps in Uganda are a luxury as they do not solve any problem unlike in Europe where property prices depend on flood risk assessments as captured in the quote below:

"In the end, we do not need flood risk maps. You can only solve a problem that you have. Flood risk maps are a luxury, and they don't solve any problem." [Consultant 2, Key Informant Interviews]

That said, however, flood risk mapping has gained attention in the last decade [12, 36] despite data insufficiency that characterises the SSA. Nonetheless, key debates on understanding key context-specific urban conditions have been invigorated to tackle the increasing cases of urban flooding in African cities.

3.5 Socio-economic flood risk drivers

The current urban geomorphology in African cities takes its shape from its colonial history. It is widely acknowledged that colonialism fostered socially excluded settlements that resulted in the urban divide between high-end, formally planned suburbs for the rich and unplanned settlements for the poor. However, as cities including Kampala continue to develop, social exclusion is still visible although it is no longer blamed solely on colonial urban planning and development codes but rather on urban practice-based realities, inequitable distribution of government resources and urban politics. This observation is based on findings from both key informant interviews and the web survey supported by geographic information system (GIS) mapping. With more than 60% of Kampala's urban population residing in vast, interconnected informal settlements, flood risks are exacerbated along the lines of poor urban planning, inadequate housing facilities and oppressive policies. To derive valuable insights into the intricate dynamics of urban flooding, Kampala's flood hotspots are examined whilst considering the socio-economic factors at play.

According to GIS spatial analyses, more than 1/3 of slum settlements across Kampala form a significant part of the flood hotspots and are affected by flooding perennially (see **Figure 6**). This is consistent with the results of the web-based survey findings and in-depth interviews that highlighted an uncontrolled incursion of the urban built environment within the flood hotspots across Kampala. What is more interesting in both the web-based survey and in-depth interview findings is that informal settlements were categorised among the significant flood drivers in Kampala after wetland loss, poor waste management, inadequate drainage infrastructure and weak enforcement of development controls.

Slum settlements not only are drivers of urban flooding but also suffer the consequences of their own existence. This overlapping contrast introduces a level of

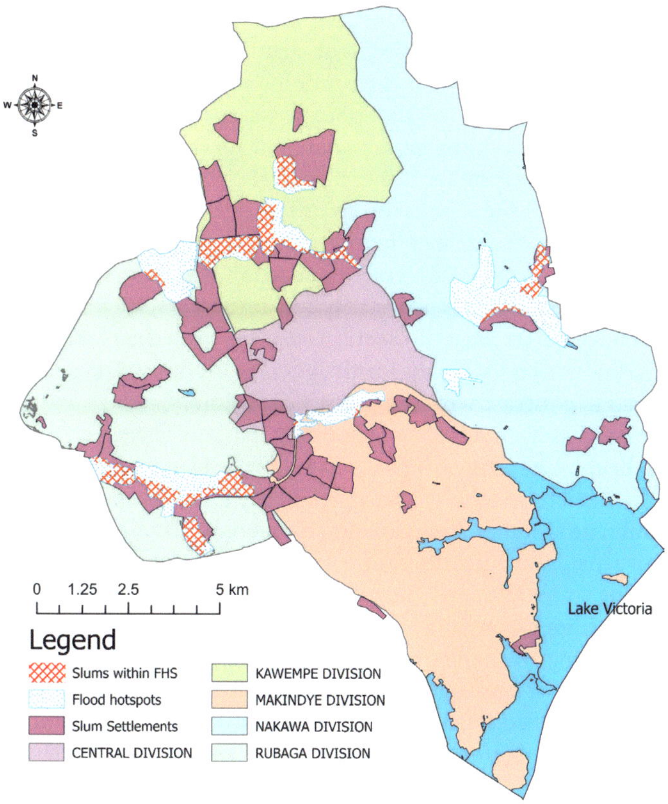

Figure 6.
Shows the overlap of flood hotspots and informal settlements (source: Author, 2022).

complexity that requires a detailed unpacking if flood hazard vulnerability in particular slum settlements is to be better understood. To trace the overlapping contrast surrounding flood-prone slum settlements in Kampala, flood hotspots across the five divisions were examined. One of the key challenges associated with informal settlements is the issue of poor waste management. KCCA points out that poor waste disposal in slums and generally across the city interferes with the normal operations of drainage networks, thereby causing flooding. The growing problem of solid waste management remains a neglected aspect of urban management, which poses an increasingly serious issue in respect of flood risk management. In Kampala, poor solid waste management was identified through interviews and the web-based survey as one of the key socio-economic and institutional drivers that amplified flooding within slum settlements. As the population grows, proper waste management degenerates among the urban poor who cannot afford private waste collection services.

KCCA cannot handle all the waste generated across the city promptly, much as it was inaugurated to improve the image of the city through proper waste collection and management, which had failed in the former regime. Whilst there is a consensus to suggest that garbage collection has increased considerably from 30% in 2010 to more than 60% in 2019 under the new KCCA leadership [37], the findings of this research indicate that poor waste disposal remains among the topmost-ranked flood risk drivers across the hotspots. Across other African cities, heaps of poorly disposed waste characterise the urban landscape, particularly in informal settlements [3, 23].

The challenge is that the urban poor in slums rely on KCCA services, which are described as 'inconsistent' and 'inconvenient' according to the web-based survey findings. Nonetheless, whilst 41% of the participants believe there has been a slight change for the better, 31% reported that poor waste management is getting worse and ultimately worsening the flooding situation whenever it rains. The remaining 28% of the web-based survey respondents were neutral citing a 'no change' status in solid waste management across the city in the last 10 years of KCCA leadership. Perhaps, the disparity in views about waste collection and management among the respondents reflects the differences in the service operation whereby, it is widely believed that proper waste management is not evenly felt across the city scale. The affluent suburbs seem to have better waste collection services than the rest of the other areas. Ultimately, KCCA is generally hailed for the good work done as far as waste management is concerned in contrast with the past KCC regime.

Poor solid waste management affects flooding in many ways. For this research, it was found that the most common impact related to the obstruction of waterways and drainage channels, whereby solid wastes particularly garbage, plastics and polythene bags clog the drainage systems. This clogging happens especially at pinch points such as insufficiently dimensioned culverts, bridges or other constrictions along the drainage channels, impeding water flow and, hence, flooding. Solid wastes find their way into drainage channels either through (a) irresponsible dumping of wastes in open drainage channels or (b) by being washed or blown into the channels by the stormwater or wind. The two characterise poor waste disposal tendencies that are common in Kampala.

Both regular waste collection and maintenance of the drainage channels would work better in tandem for better flood management results. The challenge is said to be that slums make it impossible to enforce regular waste collection and drainage channel maintenance schedules due to lack of access [38]. Whether through habitual poor waste disposal or low levels of waste collection coverage, the impact on urban flooding from wastes has been closely associated with poorly planned settlements in many African cities, that is, in Port Harcourt, Nigeria [14, 23]; Dar es Salaam, Tanzania [29] and Accra, Ghana [3]. However, for Kampala, there are counterclaims to this. Based on the findings, it was pointed out that (a) there is laxity in handling the garbage and silt dredged out of the drains and (b) there is impromptu arrival of garbage collection trucks and low maintenance of drainage channels to remove suspended solids. In the end, the garbage piles along the drainage channel banks only to be washed back during precipitation.

4. Conclusion

This chapter uses multiple qualitative methods to reveal the socio-spatial contexts that exacerbate urban flooding in a typical SSA city using Kampala's case. GIS was used as part of the methodology to alienate flood hotspots and to better understand complex contextual relationships that are usually overlooked when considering flood risks in a rapidly changing urban environment. For Kampala, contemporary flood drivers, that is, the habitation of flood plains, incursion of informal settlements and deficits in flood control infrastructure as identified through this research, owe their origins to different context-specific urban conditions, processes and interactions which are similar to other SSA cities. For instance, whilst floodplain habitation and urban flooding remain a global phenomenon due to rapid urbanisation and informal settlements in SSA cities [2, 33], the findings of this chapter indicate that Kampala's floodplain habitation has its genesis primarily rooted in poor planning for urban space and resources such as wetlands. The encroachment on wetlands restricts the space available for conveying or storing stormwater, hence causing frequent severe flooding within these areas.

That said, the factors behind the habitation of flood-sensitive zones in SSA are diversified depending on individual case studies [2, 3, 28]. It remains widely accepted that the lack of stricter urban planning and development codes has (re)produced specific urban settings that are susceptible to urban flood risk. Specifically, the present study established that urban flooding in informal settlements is not generic across African cities but rather with respect to context-specific aspects of the slum conditions, which may vary from one area to another. This is anchored on GIS findings in which it is revealed that more than 1/3 (out of 57) of informal settlements are found in flood hotspots. That said, the existing poorly planned and underdeveloped drainage systems double as dumping sites in most slum settlements. This leads to blockage of the limited drainage space, hence causing frequent localised flooding within respective areas. Improving drainage infrastructure would demand an improved system that checks indiscriminate waste disposal. Essentially, this goes beyond the engineering departments to include a wide range of urban actors and government agencies.

Ultimately, the widespread informal settlements directly reduce the wetland surface area and further create impermeable surfaces that prevent stormwater from soaking away. This influences widespread flooding but to varying extents. The backfilling of wetlands from such activities diverts stormwater to flow into new channels or local drainage systems, thereby creating new flooding pinch points. The informal settlements encroach on primary drainage channel reserves and generate high runoff volumes from the hard surfaces, leading to sporadic flooding due to insufficient drainage channels. Such activities physically reduce the wetland surface and inevitably increase impermeable hard surfaces that amplify stormwater runoff generation, hence flooding. Retrofitting bigger culverts may be possible, although this does not solve the fundamental underlying issue of loss of porous surfaces and the absorptive capacity for the wetland.

This research further demonstrated that the posturing of government agencies and urban policy has to a large extent exacerbated flood risks in Kampala. This is because, whereas wetland encroachment has been legislated against, previous government policies legally provided legislation for industrial estates' development in wetlands and advocated for wetland reclamation as a disease-control mechanism. This polarity coupled with a lack of stricter policing has continued to subject wetlands to constant encroachment by the new migrants and the urban poor. On the other hand, government efforts towards effective flood management regimes are concentrated on desilting storm drains, construction of drainage channels, demolition and forced evictions. These confrontational, inconsistent and non-participatory government approaches have not been saucerful in attenuating flood risks in other SSA cities [3, 33]. Ultimately, the solution would be to deal with the unresolved issue of wetland loss, which is almost impossible given the occupancy level of the urban poor.

Acknowledgements

This work was fully supported by the University of Manchester through the School of Environment, Education and Development (SEED Scholarship) awarded to the author in 2018/2019. The author also acknowledges the constructive insights from Prof Graham Haughton and Dr. Ransford Antwi Acheampong during the supervision of this research.

Parts of this chapter were previously published in the doctoral thesis by the same author: Denis Arinabo, Understanding the evolving nature of urban flooding in Sub-Saharan Africa: Empirical evidence from Kampala city, Uganda, 2023 [Unpublished doctoral thesis, University of Manchester]. Available from: https://pure.manchester.ac.uk/ws/portalfiles/portal/295565836/FULL_TEXT.PDF

Conflict of interest

The author declares no conflict of interest.

Author details

Denis Arinabo
School of Environment, Education and Development, The University of Manchester, Manchester, United Kingdom

*Address all correspondence to: denis.arinabo@manchester.ac.uk; den.arinabo@gmail.com

References

[1] Li L, Uyttenhove P, Van Eetvelde V. Planning green infrastructure to mitigate urban surface water flooding risk—A methodology to identify priority areas applied in the city of Ghent. Landscape and Urban Planning. 2020;**194**:103703

[2] Jha AK, Bloch R, Lamond J. Cities and Flooding: A Guide to Integrated Urban Flood Risk Management for the 21st Century. Washington, DC: World Bank; 2012. Available from: https://openknowledge.worldbank.org/handle/10986/2241. License: CC BY 3.0 IGO

[3] Amoako C, Cobbinah PB, Darkwah RM. Complex twist of fate: The geopolitics of flood management regimes in Accra, Ghana. Cities. 2019;**89**:209-217

[4] Santos PP, Pereira S, Zêzere JL, Tavares AO, Reis E, Garcia RA, et al. A comprehensive approach to understanding flood risk drivers at the municipal level. Journal of Environmental Management. 2020;**260**:110127

[5] Tiepolo M. Flood risk reduction and climate change in large cities south of the Sahara. In: Climate Change Vulnerability in Southern African Cities. Cham: Springer; 2014. pp. 19-36

[6] Otis J. After 11 months Colombia asks, who'll stop the rain? Time Magazine. 2011, May 10

[7] Morales L. Colombia's flooded economy. Pulitzer Center on Crisis Reporting. 2011. Available from: https://pulitzercenter.org/articles/colombia-bogota-rain-flood-wetlands-ecosystem?format=print

[8] WHO International. Floods; 2020. Available from: https://www.who.int/health-topics/floods#tab=tab_1 [Accessed: 3 March 2024]

[9] OCHA. United Nations Office for the Coordination of Humanitarian Affairs—West and Central Africa: Flooding Situation As of 23 October 2020. New York City: Relief Web; 2020. Available from: https://reliefweb.int/report/niger/west-and-central-africa-flooding-situation-23-october-2020 [Accessed: 3 March 2024]

[10] Tripathee A, Boyrie N. International Federation of Red Cross and Red Crescent Societies (IFRC) Information Bulletin—Africa: Multiple Disasters and Hazards. New York City: Relief Web; 2020. Available from: https://reliefweb.int/report/world/ifrc-information-bulletin-africa-multiple-disasters-and-hazards [Accessed: 3 March 2022]

[11] Fraser A, Leck H, Parnell S, Pelling M. Africa's urban risk and resilience. International Journal of Disaster Risk Reduction. 2017;**26**:1-6

[12] Arinabo D. Unveiling the role of contextual factors in the evolution of urban floods in Sub-Saharan Africa: Lessons from Kampala city. Environmental Science & Policy. 2022;**137**:239-248

[13] Doocy S, Daniels A, Murray S, Kirsch TD. The human impact of floods: A historical review of events 1980-2009 and systematic literature review. PLoS Currents. 2013;**5**

[14] Echendu AJ. Human factors vs climate change; experts' view of drivers of flooding in Nigeria. Natural Hazards Research. 2023a;**3**(2):240-246

[15] He Y, Thies S, Avner P, Rentschler J. Flood impacts on urban transit and accessibility—A case study of Kinshasa. Transportation Research

Part D: Transport and Environment. 2021;**96**:102889

[16] Ziervogel G, Pelling M, Cartwright A, Chu E, Deshpande T, Harris L, et al. Inserting rights and justice into urban resilience: A focus on everyday risk. Environment and Urbanization. 2017;**29**(1):123-138

[17] Goodfellow T. Planning and development regulation amid rapid urban growth: Explaining divergent trajectories in Africa. Geoforum. 2013;**48**:83-93

[18] Macchi S. Adaptation to incremental climate stress in urban regions: Tailoring an approach to large cities in Sub-Saharan Africa. In: Climate Change Vulnerability in Southern African Cities. Cham: Springer; 2014. pp. 3-18

[19] IPCC. Climate Change 2014: Impacts, Adaptation, and Vulnerability. Geneva, Switzerland: IPCC; 2014. Available from: http://www.Ipcc.Ch/report/ar5/wg2/ [Accessed 22 February 2022]

[20] Khaki M, Awange J. The 2019-2020 rise in Lake Victoria monitored from space: Exploiting the state-of-the-art GRACE-FO and the newly released ERA-5 reanalysis products. Sensors. 2021;**21**(13):4304

[21] Mugume SN, Abasabyoona G, Engwau J, Sempewo J, Van de Sande B, Butler D. Assessment of the impact of the rise in Lake Victoria water levels on urban flooding using a GIS-based spatial flood modelling approach. Urban Water Journal. 2024;**21**(2):219-233

[22] Arinabo D. Reconciling multiple forms of flood risk knowledge and institutional responses: Insights from Kampala's flood management regime. International Journal of Disaster Risk Reduction. 2023;**94**:103829

[23] Echendu AJ. Flooding and waste disposal practices of urban residents in Nigeria. GeoHazards. 2023b;**4**(4):350-366

[24] Satterthwaite D, Mitlin D, Bartlett S. Is it possible to reach low-income urban dwellers with good-quality sanitation? Environment and Urbanization. 2015;**27**(1):3-18

[25] Richmond A, Myers I, Namuli H. Urban informality and vulnerability: A case study in Kampala, Uganda. Urban Science. 2018;**2**(1):22

[26] Amoako C, Frimpong Boamah E. Becoming vulnerable to flooding: An urban assemblage view of flooding in an African City. Planning Theory & Practice. 2020;**21**(3):371-391

[27] Echendu AJ. Relationship between urban planning and flooding in Port Harcourt city, Nigeria; insights from planning professionals. Journal of Flood Risk Management. 2021;**14**(2):e12693

[28] Poku-Boansi M, Amoako C, Owusu-Ansah JK, Cobbinah PB. What the state does but fails: Exploring smart options for urban flood risk management in informal Accra, Ghana. City and Environment Interactions. 2020;**5**:100038

[29] Sakijege T, Dakyaga F. Going beyond generalisation: Perspective on the persistence of urban floods in Dar es Salaam. Natural Hazards. 2023;**115**(3):1909-1926

[30] Twinomuhangi R, Sseviiri H, Mulinde C, Mukwaya PI, Nimusiima A, Kato AM. Perceptions and vulnerability to climate change among the urban poor in Kampala City, Uganda. Regional Environmental Change. 2021;**21**:1-13

[31] Isunju JB, Kemp J. Spatiotemporal analysis of encroachment on wetlands:

A case of Nakivubo wetland in Kampala, Uganda. Environmental Monitoring and Assessment. 2016;**188**(4):203

[32] Kalabamu FT. Land tenure reforms and persistence of land conflicts in Sub-Saharan Africa–the case of Botswana. Land Use Policy. 2019;**81**:337-345

[33] Douglas I. The challenge of urban poverty for the use of green infrastructure on floodplains and wetlands to reduce flood impacts in intertropical Africa. Landscape and Urban Planning. 2016;**180**:262-272

[34] Mguni P, Herslund L, Jensen MB. Sustainable urban drainage systems: Examining the potential for green infrastructure-based stormwater management for Sub-Saharan cities. Natural Hazards. 2016;**82**(2):241-257

[35] Haughton G, White I. Risky spaces: Creating, contesting and communicating lines on environmental hazard maps. Transactions of the Institute of British Geographers. 2018;**43**(3):435-448

[36] Nkwunonwo UC, Whitworth M, Baily B. A review and critical analysis of the efforts towards urban flood risk management in the Lagos region of Nigeria. Natural Hazards and Earth System Sciences. 2016;**16**(2):349-369

[37] Aryampa S, Maheshwari B, Sabiiti E, Bateganya NL, Bukenya B. Status of waste management in the East African Cities: Understanding the drivers of waste generation, collection and disposal and their impacts on Kampala City's sustainability. Sustainability. 2019;**11**(19):5523

[38] KCCA. [ebook]; 2015. Available from: https://www.kcca.go.ug/uploads/KCCA_STRATEGI_PLAN_2015-2016.pdf. [Accessed: 3 October 2023]

Chapter 4

Real Options Analysis for Flood Control Facilities Taking Account of Climate Change Uncertainty

Shunsuke Segi, Katsumi Wakigawa, Toshiyuki Nishikori and Yusuke Kobayashi

Abstract

As global warming progresses, predictably the amount of rainfall and the frequency of heavy rainfall will increase. Therefore, it is necessary to develop flood control facilities to deal with the increase in rainfall. However, there is a great deal of uncertainty about how quickly climate change will progress and how much the temperature will rise. Thus, it is necessary to take into consideration every scenario related to future increase in rainfall so as to properly determine the timing of construction of flood control facilities. This study proposes a real options analysis (ROA) model to control the budget allocation for flood control facilities, considering the uncertainty of climate change, along with a heuristic solution method for it. Furthermore, we apply the proposed ROA model and solution method to a basin in Japan. The result confirms that the optimal solution of the ROA model significantly reduces the social cost compared to inflexible planning.

Keywords: climate change uncertainty, flood management, real options analysis, dynamic optimization, budget allocation

1. Introduction

As global warming progresses, it is predicted that the amount of rainfall and the frequency of heavy rainfall will increase. Therefore, it is necessary to develop flood control facilities to deal with the increasing rainfall. However, it is considerably uncertain how quickly climate change will progress and how much temperature will rise. Hence, it is challenging to properly determine the timing of construction of flood control facilities by assuming a deterministic magnitude of rainfall. Therefore, an analytical framework that takes into consideration every scenario related to future temperature changes and the associated increase in rainfall becomes necessary to properly allocate limited budgets to flood control facilities.

This study proposes an ROA model that optimizes budget allocation to flood control facilities while considering the uncertainty of climate change. ROA is a decision-making method used when there is uncertainty in the future environment. ROA can optimize situation-dependent decision-making according to future scenarios.

IntechOpen

The proposed ROA model utilizes the results of runoff and flood analysis assuming various magnitudes of rainfalls and hyetographs as its input data. Based on these, the relation between the increases in rainfall, the development of flood control facilities, and the flood damage in the river basin are modeled as piecewise linear functions. The modeling framework allows us to apply the ROA model to real basins.

Solving the proposed ROA model is challenging. Therefore, this study proposes a heuristic solution method that applies the genetic algorithm. Moreover, this study implements the proposed ROA model and solution method to a river basin in Japan to verify its effectiveness.

The structure of this chapter is outlined as follows. Section 2 provides a review of existing research and presents the position of this study. Section 3 formularizes the ROA model. The solution method to this problem is proposed in Section 4. Finally, Section 5 applies the proposed ROA model and the solution method to a river basin in Japan to verify their effectiveness.

2. Review

The influence of global warming on future flood hazards, such as the rainfall and river discharge, is being studied with the help of climate projections [1–3]. While these studies analyzed the flood hazard in a given climate change scenario (e.g., RCP8.5 scenario), the uncertainties regarding the climate change scenario itself were not analyzed. Thus, no direct answer to the question of how various types of flood control facilities should be developed in response to the uncertain progression of global warming is obtained.

The method for flood risk management that considers the uncertainty of the climate change process is also a subject of increasing interest. The most widely used method is to select a specific scenario of the climate change and conduct an evaluation under that scenario. For example, Rojas et al. [4] used the ensemble data from climate simulations under a certain scenario to analyze how flood damage increases as climate change progresses. Kind [5] and Eijgenraam et al. [6] selected a scenario wherein the flood occurrence probability increases with climate change and analyzed a dynamic optimization problem of embankment height.

Recently, there have been more sophisticated evaluation methods employing the framework of ROA [7, 8]. These studies comprehensively considered the possible future scenarios to perform evaluations of the optimal development timing for flood control facilities. To respond to the uncertainties of future climate change, it is desirable to prepare a plan that considers full range of scenarios in advance and execute the development of flood control facilities adaptively in coordination with the climate change monitoring results. ROA can be utilized as the framework to optimize such an adaptive plan.

Most existing studies about ROA in the field of flood risk management are limited to virtual/conceptual analysis. The exception is the study of Kind et al. [8], who performed analyses of real rivers. Kind et al. set a realistic future scenario and then optimized the design of the embankment height for the Rhine tributaries in the Netherlands.

The studies mentioned above all presumed a situation where there is no upper limit to the government budget that can be allocated to the development of flood control facilities. Although it is useful for examining the normative amount of budget to be allocated to the development of flood control facilities, applying the framework of these studies to environments where there are budgetary limits is difficult. Therefore, our study assumes a situation wherein there is an upper limit to the single fiscal year budget that can be allocated to the development of flood control facilities and proposes an ROA model that dynamically optimizes the budget allocation to flood control facilities under climate change uncertainty. This study is novel as it enables more practical analysis through applying such a framework.

3. Model

3.1 Climate change scenario

Time is discretized into multiple periods expressed as $t = 0, 1, 2, \cdots$. $t = 0$ indicates the current time. $t = T$ indicates the final period considered in a climate change scenario. This study sets $t = 0$ to be 2020 and $t = T = 80$ at 2100. The climate change scenario of $0 \leq t \leq 80$ is modeled.

This study uses the data obtained from the graph published in the report of Intergovernmental Panel on Climate Change [9] for setting the stochastic process that the future temperature change (s) will follow. The blue and red lines in **Figure 1** are the graphical representations of the read data (the broken lines will be discussed later). The horizontal axis represents the year, and the vertical axis represents temperature change s[°C]. The blue and red lines are the average of the prediction models under RCP2.6 scenario and RCP8.5 scenario, respectively. Herein, it is postulated that the region between these curves represents the 90% confidence interval of the future temperature change (s) prediction. Moreover, the logarithm of s is postulated to follow the random walk with drift.

$$\ln s_t = \ln s_{t-1} + \mu_t + \sigma_t \varepsilon_t \tag{1}$$

Here, s_t[°C] represents the temperature change at period t. μ_t and σ_t are the parameters of the normal distribution $N(\mu_t, \sigma_t^2)$ to which the rate of change between

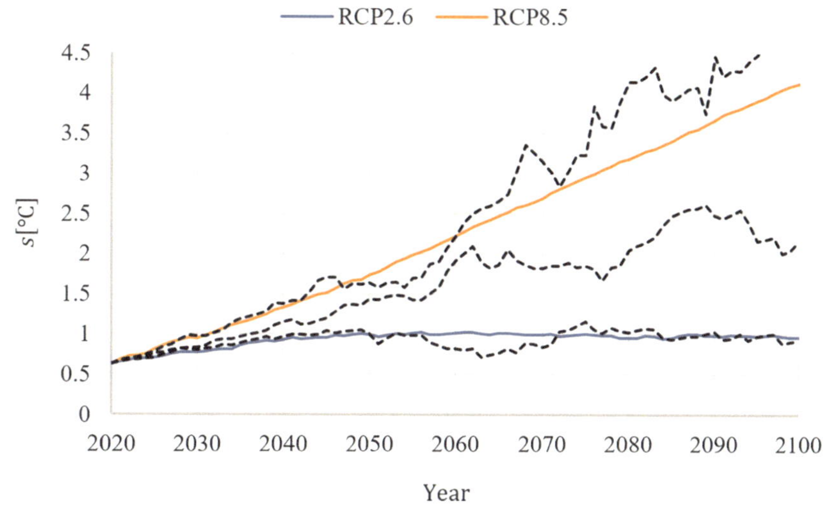

Figure 1.
Future climate change scenarios.

period $t-1$ and period t follows. The values of μ_t and σ_t differ at each period. The interval between each period is set at one year. ε_t is the stochastic variable that follows the standard normal distribution.

Under the assumption of Eq. (1), the expected value and variance of $\ln s_t$ evaluated at period 0 are expressed respectively as follows:

$$\mathrm{E}_0[\ln s_t] = \ln s_0 + \sum_{\tau=1}^{t} \mu_\tau \tag{2}$$

$$\mathrm{V}_0[\ln s_t] = \sum_{\tau=1}^{t} \sigma_\tau^2 \tag{3}$$

The temperature change s_0 at period 0 is given exogenously. By using $\mathrm{E}_0[\ln s_t]$ and $\mathrm{V}_0[\ln s_t]$, the 90% confidence interval of $\ln s_t$ predicted at period 0 can be expressed as follows:

$$\mathrm{E}_0[\ln s_t] - 1.64\sqrt{\mathrm{V}_0[\ln s_t]} \le \ln s_t \le \mathrm{E}_0[\ln s_t] + 1.64\sqrt{\mathrm{V}_0[\ln s_t]} \tag{4}$$

The lower limit of this confidence interval is set to $\ln s_t^{2.6}$, while its upper limit is set to $\ln s_t^{8.5}$. Here, $s_t^{2.6}$ and $s_t^{8.5}$ are the expected temperature change at period t under the RCP2.6 and RCP8.5 scenarios, respectively. Hence, the values of the parameters μ_t and σ_t can be calibrated by using Eqs. (2)–(4) as follows:

$$\mu_t = \frac{\ln s_t^{2.6} + \ln s_t^{8.5}}{2} - \frac{\ln s_{t-1}^{2.6} + \ln s_{t-1}^{8.5}}{2} \tag{5}$$

$$\sigma_t^2 = \left[\left(\frac{\ln s_t^{8.5} - \ln s_t^{2.6}}{2 \cdot 1.64}\right)^2 - a_{t-1}\right]_+ \tag{6}$$

$$a_t = \begin{cases} \left(\dfrac{\ln s_t^{8.5} - \ln s_t^{2.6}}{2 \cdot 1.64}\right)^2 & \text{if } \left(\dfrac{\ln s_t^{8.5} - \ln s_t^{2.6}}{2 \cdot 1.64}\right)^2 \ge a_{t-1} \\ a_{t-1} & \text{otherwise} \end{cases} \tag{7}$$

$[x]_+$ is the symbol that represents $\max[x, 0]$; when the value of x is negative, it indicates rounding up to 0. a_t is a supplementary variable. In most of the cases, σ_t^2 is calibrated as the difference of $\left[\left(\ln s_t^{8.5} - \ln s_t^{2.6}\right)(2 \cdot 1.64)\right]^2$ between successive periods. However, this value can be negative. Eqs. (6) and (7) can be applicable to such a case.

The three broken lines in **Figure 1** are examples of sample paths generated by using Eq. (1) with the calibrated μ_t and σ_t. As the sample paths show, the random walk model enables us to consider diverse climate change scenarios.

3.2 Relationship between temperature change and amount of rainfall

We model the relationship between the temperature change and the annual maximum 12-hour rainfall r[mm] in the upstream of a basin. Let r_t denote the annual maximum rainfall at period t, which is a stochastic variable. The stochastic distribution that r_t follows is expressed as a Gumbel distribution and its cumulative distribution function Φ is formularized as follows:

$$\Phi(r_t|s_t) = \exp\left[-e^{-\frac{r_t-\theta(s_t)}{\eta(s_t)}}\right] \tag{8}$$

$\theta(s_t)$ and $\eta(s_t)$ are the parameters that determine the shape of the Gumbel distribution and are dependent on the temperature change s_t at period t. These parameters are formularized as follows:

$$\theta(s_t) = (1 + us_t)\theta_{\text{base}} \tag{9}$$

$$\eta(s_t) = (1 + us_t)\eta_{\text{base}} \tag{10}$$

Here, θ_{base} and η_{base}, respectively, represent the parameters of the Gumbel distribution in an environment where the temperature change is $s = 0$. u is a constant parameter. Eqs. (9) and (10) indicate that when the temperature increases by one degree, the expected value and the standard deviation of the annual maximum rainfall will increase by $1 + u$ times. This implies that the risk of heavy rainfall increases as the temperature rises. The value of u is set referring to the climate projections.

3.3 Consideration of several hyetographs

We consider N_W shapes of hyetographs (space–time distribution of rainfall). Each shape of hyetograph realizes with equal probability $(1/N_W)$. $w_t \in \{1, \cdots, N_W\}$ represents the shape of hyetograph at period t, which is a stochastic variable.

At each period, the 12-hour rainfall r_t and the shape of hyetograph w_t are determined stochastically. The hyetograph is scaled so that the 12-hour rainfall calculated from it coincides with r_t. Given the scaled hyetograph, the flood damage at period t can be determined by simulation analysis.

3.4 Evaluation method of flood damage

To solve the ROA model of this study, it is necessary to evaluate the flood damage in extremely diverse situations, where the amount of rainfall, hyetographs (space–time distribution of rainfall), the location of levee breaches, and the presence of flood control facilities differ. However, it is difficult to conduct runoff and flood analysis for all the situations based on a hydrology model due to the required calculation time. Therefore, the flood damage is approximately evaluated using the method presented in this subsection.

First, several conditions with different amounts of rainfall, hyetographs, location of levee breaches and the presence of flood control facilities are set, and hydrology-based runoff/flood analyses are performed. Next, the results of hydrology-based runoff/flood analyses are used to construct piecewise linear functions, which contain the information required for flood damage evaluation, such as the relation between the rainfall/hyetograph and the peak discharge. These piecewise linear functions are used to approximately evaluate the flood damage under various situations when performing the ROA.

The following is the method to approximately evaluate flood damage using the results of runoff/flood analysis in a situation where the 12-hour rainfall r_t and the shape of hyetograph w_t realize.

We divide the land in the target basin into several residential areas. The number of the residential areas is represented as N_J. Individual residential areas are indicated by the subscript $j \in \{1, 2, \cdots, N_J\}$. Different residential areas are set at the left and the right bank of the river. Each residential area has one candidate point where a levee breach could occur (levee breach candidate point). The area number j is organized in such a way that the levee breach points in the upstream areas come before those in the downstream areas.

Let q_{tj}[m³/s] denote the peak discharge at the levee breach point of area j at period t. q_{tj} is evaluated using the following equation:

$$q_{tj} = K_j(r_t|w_t) - \sum_{i=1}^{N_I} \Delta_{ij}^X(r_t|w_t) f_i(x_{ti}) - \sum_{j'=1}^{j-1} \Delta_{tj'j}^B \tag{11}$$

The components of this equation are explained below.

$K_j(r|w)$ is the piecewise linear function for evaluating the peak discharge q_{tj} of area j when hyetograph w and rainfall r are realized under the present conditions (beginning of the period $t = 0$), and it is constructed by using the result of runoff analysis. The construction procedure is as follows. First, N_R number(s) of water levels $r^1, r^2, \cdots, r^{N_R}$ are set as the 12-hour rainfall. For example, we can use the 12-hour rainfall for return periods of 10, 20, 50, 100, ... years for them. These rainfalls are ordered such that $r^1 < r^2 < \cdots < r^{N_R}$ is valid. Subsequently, runoff analysis is performed under the present conditions with the hyetograph w and rainfall r^k $(1 \le k \le N_R)$ to obtain the peak discharge q_j^k of each area. No levee breach is assumed to occur during this analysis. A point $\left(r^k, q_j^{w,k}\right)$ $(1 \le k \le N_R)$ on the r–q_j plane will be obtained as a result of the runoff analysis. By connecting these points with straight lines, the piecewise linear function $K_j(r|w)$ is constructed.

$$K_j(r|w) = \mathrm{PL}\left(\left\{r^k, q_j^{w,k}\right\}_{k=1}^{N_R}\right) \tag{12}$$

$\mathrm{PL}\left(\left\{x^k, y^k\right\}_{k=1}^{N}\right)$ is a piecewise linear function obtained by connecting N numbers of points (x^k, y^k) $(1 \le k \le N)$ with straight lines and is defined as follows:

$$\mathrm{PL}\left(\left\{x^k, y^k\right\}_{k=1}^{N}\right) = \begin{cases} \left[y^k + \dfrac{y^{k+1} - y^k}{x^{k+1} - x^k} \cdot \left(x - x^k\right)\right]_+ & \left(x \in \left[x^k, x^{k+1}\right]\right) \\ \left[y^1 + \dfrac{y^2 - y^1}{x^2 - x^1} \cdot \left(x - x^1\right)\right]_+ & \left(x \in \left[0, x^1\right]\right) \\ \left[y^{N-1} + \dfrac{y^N - y^{N-1}}{x^N - x^{N-1}} \cdot \left(x - x^{N-1}\right)\right]_+ & \left(x \in \left[x^N, \infty\right]\right) \end{cases} \tag{13}$$

When $x < x^1$, the straight line that connects (x^1, y^1) and (x^2, y^2) is extrapolated. When $x > x^N$, the straight line that connects $\left(x^{N-1}, y^{N-1}\right)$ and $\left(x^N, y^N\right)$ is extrapolated. When the extrapolated straight line is negative, its value is rounded up to 0.

The graphs shown in **Figure 2** illustrate examples of $K_j(r|w)$ used in the analysis of Section 5. These four graphs of $K_j(r|w)$ demonstrates the relation between rainfall r and the peak discharge q of a given area. Each graph corresponds to a specific hyetograph w. Each marker on the graph is a (r, q) pair presenting the result of a runoff analysis.

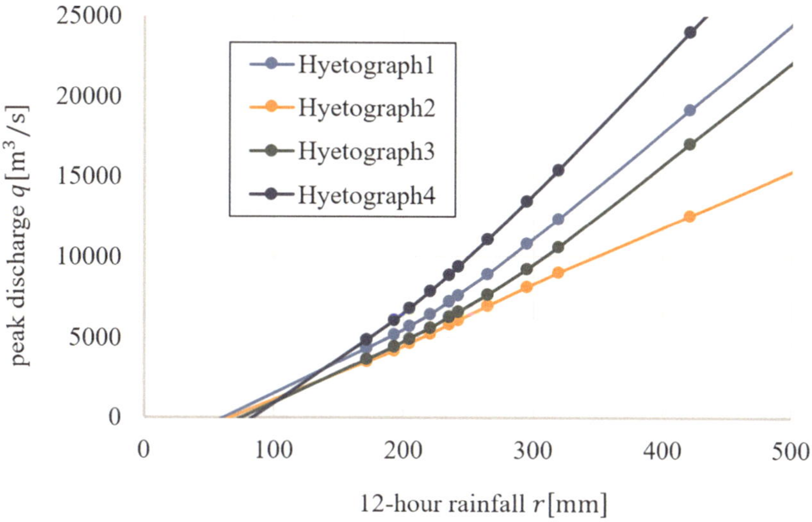

Figure 2.
Examples of $K_j(r|w)$.

There are N_I number of flood control facilities that have yet to be constructed. Each of these facilities is represented by the subscript $i \in \{1, 2, \cdots, N_J\}$. $\Delta_{ij}^{X}(r|w)$ is the function for evaluating the reduction effect of the development of flood control facility i on the peak discharge of area j and is constructed by employing the results of the runoff analysis. The construction procedure is as follows. Consider a situation where the only change made to the current river channel is the construction of flood control facility i (i.e., no other flood control facilities are developed). Runoff analysis is performed under this condition with hyetograph w and rainfall r^k $(1 \le k \le N_R)$ to obtain the pair $\left(r^k, \Delta q_{ij}^{w,k}\right)$ of rainfall r^k and the reduction amount of the peak discharge of area j ($q_j^{w,k}$ is used for the reference to calculate the reduction amount). By connecting these points with straight lines, the piecewise linear function $\Delta_{ij}^{X}(r|w)$ is obtained.

$$\Delta_{ij}^{X}(r|w) = \mathrm{PL}\left(\left\{r^k, \Delta q_{ij}^{w,k}\right\}_{k=1}^{N_R}\right) \tag{14}$$

The graphs shown in **Figure 3** illustrate an example of $\Delta_{ij}^{X}(r|w)$ used in the analysis in Section 5. These graphs depict the amount of reduction in the peak flow of a given area when a dam is constructed in the upstream. The graphs do not show a monotonic increase because they include the situation when the dam has reached its capacity.

$x_{ti} \in [0, 1]$ describes the progress rate of the construction of flood control facility i at the beginning of period t. $x_{ti} = 0$ indicates that its construction has not been started, and $x_{ti} = 1$ indicates that it has been completed. $f_i(x_i)$ is the function showing the level of the effect of flood control facility i when its construction progress rate is x_i. When the effect of facility i is zero until construction is completed, such as a dam, $f_i(x_i)$ is expressed as follows.

$$f_i(x_i) = \begin{cases} 0 & (0 \le x_i < 1) \\ 1 & (x_i = 1) \end{cases} \quad (i \in I_1) \tag{15}$$

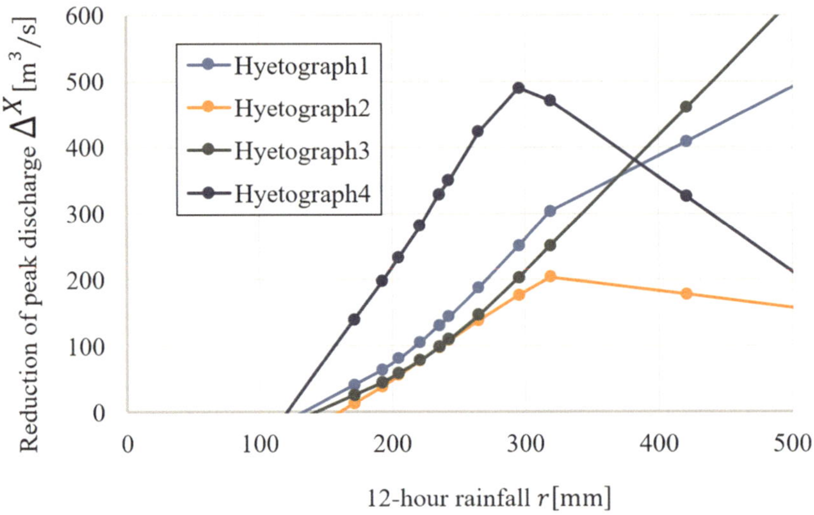

Figure 3.
Examples of $\Delta_{tj}^X(r|w)$.

I_1 denotes the set of the facilities of such type. Meanwhile, when the effect of facility i gradually manifests as construction progresses, even if construction is not completed, such as riverbed deepening (we regard deepened riverbed as a kind of facility), $f_i(x_i)$ is expressed as follows.

$$f_i(x_i) = x_i \ (i \in I_2) \tag{16}$$

I_2 denotes the set of the facilities of such type.

It is assumed that the reduction effects in the peak discharge due to the construction of flood control facilities sum up linearly. Under this assumption, the second term on the right side of Eq. (11) depicts the reduction effect of the flood control facilities at period t on the peak discharge of area j.

$\Delta_{tj'j}^B$ indicates the peak flow reduction effect at downstream area j when a levee breach occurs in area j'. $\Delta_{tj'j}^B$ is stochastically determined following the process discussed below.

The following is the process for determining the economic damage in the basin with Eq. (11). First, we set $j = 1$. When $j = 1$, the third term on the right side of Eq. (11) does not exist. Therefore, given r_t and w_t, q_{t1} is calculated deterministically. Then, the following equation is used to determine the levee breach probability π_{tj} in area j.

$$\pi_{tj} = \begin{cases} 0 & \left(q_{tj} \le \tilde{q}_{tj}\right) \\ \dfrac{q_{tj} - \tilde{q}_{tj}}{\overline{q}_{tj} - \tilde{q}_{tj}} & \left(\tilde{q}_{tj} < q_{tj} < \overline{q}_{tj}\right) \\ 1 & \left(q_{tj} \ge \overline{q}_{tj}\right) \end{cases} \tag{17}$$

$\tilde{q}_{tj}$ is the lower limit of the peak discharge that can potentially cause a levee breach to occur in area j, and $\overline{q}_{tj}$ is the peak discharge that certainly causes a levee breach to occur. $\tilde{q}_{tj}$ and $\overline{q}_{tj}$ are affected by riverbed deepening and levee heightening. The following equations describe the impact of the development of these flood control facilities.

$$\tilde{q}_{tj} = \tilde{q}_j^{\text{base}} + \sum_{i=1}^{N_I} \overline{h}_{ij} f_i(x_{ti}) \tag{18}$$

$$\overline{q}_{tj} = \overline{q}_j^{\text{base}} + \sum_{i=1}^{N_I} \overline{h}_{ij} f_i(x_{ti}) \tag{19}$$

$\tilde{q}_j^{\text{base}}$ and $\overline{q}_j^{\text{base}}$ are the respective values of $\tilde{q}_{0j}$ and $\overline{q}_{0j}$ in the current situation. $\overline{h}_{ij}$ is the constant that expresses the effect of increasing $\tilde{q}_{tj}$ and $\overline{q}_{tj}$ due to the development of flood control facility i. If flood control facility i is either levee heightening or riverbed deepening in area j, $\tilde{q}_{tj}$ and $\overline{q}_{tj}$ increase by the same amount denoted by $\overline{h}_{ij} > 0$. For other flood control facilities, $\overline{h}_{ij} = 0$.

Let b_{tj} denote the stochastic variable that indicates the levee breach in area j at period t. If there is a levee breach, it becomes 1; if there is no breach, it becomes 0. Hence, $\Pr[b_{tj} = 1] = \pi_{tj}$. When a levee breach occurs at area j, damage is inflicted on this area. Moreover, depending on the magnitude of the flood, damage could occur in the area $j' > j$ located downstream from area j. The damage cost $d_{tjj'}$ on area j' ($j' \geq j$) caused by the levee breach in area j is determined by the following equation.

$$d_{tjj'} = b_{tj} g_{jj'} \left(q_{tj} - \sum_{i=1}^{N_I} \tilde{h}_{ij} f_i(x_{ti}) \middle| w_t \right) \tag{20}$$

$g_{jj'}(q|w)$ is the piecewise linear function that represents the economic damage incurred on area j' because of the levee breach in area j. It is postulated that economic damage $d_{tjj'}$ can be expressed as the function of the peak discharge at the levee breach point. However, if riverbed deepening has been conducted along area j, the economic damage decreases under the same peak discharge q_{tj}. Therefore, if the flood control facility i is riverbed deepening along area j, the damage is reduced by the increased river flow capacity denoted by $\tilde{h}_{ij} > 0$. For all other flood control facilities, $\tilde{h}_{ij} = 0$.

When a levee breach occurs in area j, the downstream peak discharge from this area decreases. This effect is determined by the following equation:

$$\Delta_{tjj'}^{B} = b_{tj} L_{jj'} \left(q_{tj} - \sum_{i=1}^{N_I} \tilde{h}_{ij} f_i(x_{ti}) \middle| w_t \right) \tag{21}$$

$L_{jj'}(\hat{q}|w)$ is the piecewise linear function that represents the reduction of the peak discharge in area j' when a levee breach occurs in area j ($j' > j$).

The result of runoff/flood analysis is used to construct functions $g_{jj'}(q|w)$ and $L_{jj'}(q|w)$. The construction procedure is as follows. Consider a situation where only the levee of area j breaks in the whole river channel under the present condition. Runoff/flood analysis is performed under this condition with hyetograph w and rainfall r^k ($1 \leq k \leq N_R$). This analysis provides pairs of the peak discharge of area j and the economic damage of area j' ($j' \geq j$) ($q_j^{w,k}, d_{jj'}^{w,k}$) and pairs of the peak discharge of area j and the decrease in the peak discharge of area j' ($j' > j$) ($q_j^{w,k}, \Delta q_{jj'}^{w,k}$) ($q_j^{w,k}$ is used as the reference to calculate the reduction amount). By connecting the former points with straight lines, $g_{jj'}(\hat{q}|w)$ is obtained; by connecting the latter points with straight lines, $L_{jj'}(\hat{q}|w)$ is achieved.

In this study, we assume that the land use and geographical distribution of tangible assets in the residential areas are unchanged throughout the period of time in consideration. Therefore, flood analysis is performed assuming the current land use and the function $g_{jj'}$ is not differentiated for each period.

The graphs shown in **Figures 4** and **5** are examples of $g_{jj'}(\hat{q}|w)$ and $L_{jj'}(\hat{q}|w)$ employed in the analysis of Section 5, respectively. There are sections in the graphs of $L_{jj'}(\hat{q}|w)$ that are not monotonically increasing. This is because of the increase in the

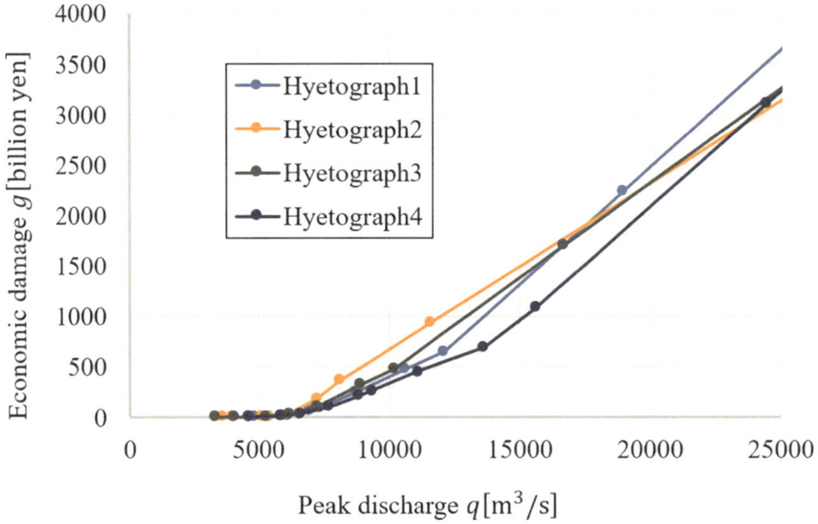

Figure 4.
Examples of $g_{jj'}(q|w)$.

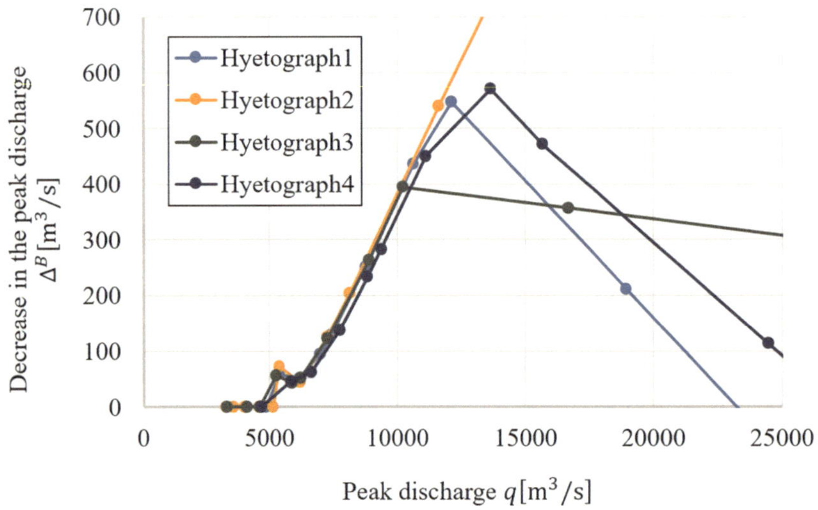

Figure 5.
Examples of $L_{jj'}(q|w)$.

levee-side water level at the breech point, which makes it difficult for flood water to flow out from the levee.

Using Eq. (11) and Eqs. (17)–(21) above, q_{t1} and π_{t1} are obtained. Moreover, the values of b_{t1}, $d_{t1j'}$ $(j' \geq 1)$, $\Delta^B_{t1j'}$and $(j' > 1)$ are stochastically achieved. Then, q_{t2} and π_{t2} are obtained using the result of $j = 1$ and Eqs. (17)–(21). Additionally, the values of b_{t2}, $d_{t2j'}$ $(j' \geq 2)$, and $\Delta^B_{t2j'}$ $(j' > 2)$ are stochastically achieved. By repeating this operation for $j = 3, \ldots, N_J$ in ascending order, the values of $d_{tjj'}$ are obtained stochastically in every area.

The economic damage d_{tj} of block j is obtained as the largest value among $d_{tj'j}$ $(j' \leq j)$.

$$d_{tj} = \max_{j' \leq j} d_{tj'j} \tag{22}$$

The reason for using this formulation is to avoid double counting the economic damage when levee breaches occur in multiple areas. Once d_{tj} of every block is determined, the total economic damage D_t of flooding at period t is evaluated as their total value.

$$D_t = \sum_{j=1}^{N_J} d_{tj} \tag{23}$$

What has been outlined in this subsection is a method to evaluate the flood damage in a situation wherein r_t and w_t are realized. Although this method interpolates and extrapolates the result of runoff/flood analysis, it can cause errors in the evaluation of economic damage. However, considering that existing studies on the application of optimization problems in the development of flood control facilities occasionally use simple functions in the estimation of economic damage caused by flood (e.g., Kind [5] and Eijgenraam et al. [6] employed simple exponential functions), the method employed in this study has a certain rationality.

3.5 Relation between budget allocation and progress of flood control facility development

Let y_{ti} denote the government expenditure spent on the construction of flood control facility i at period t. x_{ti}, which indicates the progress rate for the development of flood control facility i at the beginning of period t, changes according to y_{ti}.

$$x_{t+1,i} = x_{ti} + \frac{y_{ti}}{c_i} \tag{24}$$

c_i is the constant that represents the total construction cost of flood control facility i. When the total construction costs invested until period t reach c_i, the development of flood control facility i is complete.

y_{ti} must satisfy the following conditions:

$$y_{ti} \leq c_i(1 - x_{ti}) \tag{25}$$

$$0 \leq y_{ti} \leq \bar{y}_i \tag{26}$$

Eq. (25) implies that the progress rate $x_{t+1,i}$ does not exceed 1. Eq. (26) indicates that there is a physical upper limit to the construction processes that can be completed within one year. $\overline{y}_i$ is the constant that describes the upper limit of yearly government expenditure on flood control facility i. Eq. (26) implies that the minimum required time to build flood control facility i is $c_i/\overline{y}_i$ years.

The upper limit $\overline{Y}$ is set for the yearly budget that can be spent on the development of flood control facilities.

$$\sum_{i=1}^{N_I} y_{ti} \leq \overline{Y} \tag{27}$$

$\overline{Y}$ is given as a constant. y_{ti} is determined within the constraint of Eqs. (25)–(27). When analyzing a real basin, $\overline{Y}$ will be set based on the actual values of the yearly budget allocated to this basin. If an arbitrarily large number is used for $\overline{Y}$, one can represent the situation wherein there is no budgetary limit.

3.6 Objective function

Herein, the total expected net present value of the social cost is employed as the objective function to be minimized in the ROA model. First, social cost C_t incurred at period t is formulated as the sum of the construction costs for the flood control facilities and the total economic damage of flooding D_t.

$$C_t = \sum_{i=1}^{N_I} y_{ti} + D_t \tag{28}$$

Using C_t, the expected net present value of the social cost in terms of the monetary value at $t = 0$ is formulated as follows:

$$V_0 = \mathrm{E}_0\left[\sum_{t=0}^{T+\overline{T}} \beta^t C_t\right] \tag{29}$$

E_0 denotes the operation to calculate the expected value using the information available at the beginning of $t = 0$. β is the discount factor to convert the future monetary cost to their present value; when the annual discount rate is ρ, it is expressed as $\beta = 1/(1+\rho)$. $\overline{T}$ denotes a constant that indicates how many periods after $t \geq T$ are to be considered when evaluating the net present value of the social cost. While the final period in the consideration of climate change scenarios is $t = T$, the benefits of a flood control facility will be underestimated unless C_t $(t \geq T)$ is considered as the facility will continue to be useful even after this period. Therefore, by including the C_t of $T \leq t \leq T + \overline{T}$ periods in the objective function, underestimation will be prevented. When $t \geq T$, temperature change s_t will be maintained at temperature change s_T at period T $(s_t = s_T, t \geq \overline{T})$.

Herein, V_0 is adopted as the objective function to be minimized in the ROA model. The decision variable of the optimization problem is the government expenditure y_{ti} spent on the development of flood control facility i at each period.

3.7 Limitation of the model

The above model has several limitations. Below, we discuss the limitations and the changes to the model required to relax them. First, as mentioned earlier, the land use in the residential areas is assumed to be unchanged. Therefore, nonstructural measures using land use controls cannot be considered. In order to take such measures into consideration, it is necessary to perform flood analysis in a situation where land use is controlled and to reflect the results in the function $g_{jj'}$. Furthermore, since controlling land use imposes a burden on economic activities, it is necessary to include its social cost in C_t. Second, the model does not consider the maintenance costs that will occur after the construction of flood control facilities. In order to consider the costs, it is necessary to include them in C_t and the budget constraint of Eq. (27). Third, the model does not consider the damage caused to river ecosystems by measures such as channel deepening. In order to take such damage into account, it is necessary to include its social cost in C_t. Finally, it does not consider future changes in the construction cost of flood control facilities. Since it is difficult to predict future construction costs, consideration of multiple scenarios might be necessary.

4. Solution method of the ROA model

As the ROA model formulated in Section 3 contains many state variables, it is difficult to apply dynamic programming to it or obtain an exact solution. Thus, we propose a heuristic solution method in this section. We employ "parametric optimization" technique [10] for the solution method, wherein we formulate a policy function for determining decision variables in a form containing parameters and then to optimize these parameters through simulation.

4.1 Formulation of policy function

Owing to the structure of the problem, it will be efficient to start the development of a flood control facility when temperature change s_t exceeds a certain threshold. Based on this assumption, the following policy function is employed.

First, we define the set of flood control facilities that are candidates for budget allocation at period t, which is denoted by Λ_t. The government budget is spent on only the facilities in this set. Λ_t is determined by the following equation:

$$\Lambda_t = \{i | s_t \geq \tilde{s}_i \text{ and } x_{ti} < 1\} \cup \{i | i \in I_1 \text{ and } 0 < x_{ti} < 1\} \tag{30}$$

Here, $\tilde{s}_i$ is the unknown parameter to be set for each flood control facility i. This parameter is the threshold indicating that "if temperature change s_t exceeds this standard, the government budget should be spent to flood control facility i." The first curly brackets of Eq. (30) imply that among the flood control facilities that satisfy $s_t \geq \tilde{s}_i$, only those for which development has not been completed are included in Λ_t. With the rule of the first curly bracket alone, the development of flood control facility i started in the past would stop when temperature change s_t falls below $\tilde{s}_i$. Such an interruption of development could be potentially wasteful in case the flood control facility is included in I_1. Thus, the rule of the second curly bracket is introduced. This rule implies that the flood control facilities included in I_1 are always included in Λ_t once construction has started until it is complete. Meanwhile, as the government expenditure spent on flood control facilities included in I_2 will not become wasteful even if such development is interrupted, the rule of the second curly bracket is not applied to them.

Next, the order of priority concerning the budget allocation to the flood control facilities is determined. The priority rank of flood control facility i is denoted by rank(i). rank(i) is determined so that flood control facilities with smaller $\tilde{s}_i$ will be given higher priority.

$$\text{rank}(i) < \text{rank}(i') \Leftrightarrow \tilde{s}_i < \tilde{s}_{i'} \tag{31}$$

$$\text{rank}(i) \in \{1, 2, \cdots, N_I\} \tag{32}$$

Using Λ_t and rank(i), y_{ti} is determined as follows.

$$y_{ti} = \begin{cases} 0 & (i \notin \Lambda_t) \\ \min\left[c_i(1 - x_{ti}), \bar{y}_{ti}, \tilde{Y}_{\text{rank}(i)}\right] & (i \in \Lambda_t) \end{cases} \tag{33}$$

$$\tilde{Y}_1 = \overline{Y} \tag{34}$$

$$\tilde{Y}_{l+1} = \tilde{Y}_l - y_{t,\text{rank}^{-1}(l)} \tag{35}$$

$\tilde{Y}_l$ is the upper limit of the budget that can be allocated to the flood control facility that is lth in the list of priorities. As shown in Eq. (34), the upper limit of the budget that can be spent on the flood control facility with the first priority is equal to the annual government budget limit $\overline{Y}$. Eq. (35) indicates that budget y_{ti} is allocated in descending order of priority. rank$^{-1}(l)$ is the inverse mapping of rank(i) and represents the flood control facility i that is lth in the list of priorities. Eq. (33) implies that y_{ti} is set at the maximum amount possible within the amount satisfying the conditions of Eqs. (15) and (26) and the available budget $\tilde{Y}_{\text{rank}(i)}$. However, if flood control facility i is not included in the candidates for budget allocation, $y_{ti} = 0$.

4.2 Evaluation method of objective function

The parameters of the policy function formulated above are the threshold $\tilde{s}_i$ of the temperature change defined for each flood control facility i. By obtaining the $\tilde{s}_i$ that would minimize the objective function V_0, it is possible to approximately solve the ROA model. This is the concept of parametric optimization.

When evaluating an objective function, it is necessary to calculate the expected value. However, exact calculation of this expected value is difficult. Therefore, Monte Carlo simulation is utilized to approximately evaluate the objective function. The stochastic variables included in the model of this study are temperature change s_t, yearly maximum rainfall r_t, hyetograph w_t, and random numbers to determine the occurrence of levee breaches according to Eq. (17). After generating M numbers of sample paths for these stochastic variables, the objective function V_0 is evaluated according to the following equation.

$$V_0 \simeq \frac{1}{M} \sum_{m=1}^{M} \left[\sum_{t=0}^{T+\overline{T}} \beta^t C_t^m \right] \tag{36}$$

The subscript m $(1 \le m \le M)$ of the variable indicates the value under the simulation with mth sample path.

4.3 Optimization method of objective function

The evaluation of the objective function under the given $\tilde{s}_i$ becomes possible using Eq. (36). However, it is not possible to apply gradient-based optimization method when the objective function is evaluated in this manner. That is because the right side of Eq. (36) is a discontinuous stepped function with respect to $\tilde{s}_i$, and its gradient is zero in most of the domains. Therefore, a real-valued genetic algorithm is employed as an optimization method, which does not use the gradient of the objective function. We used the genetic algorithm presented by Yoon and Kim [11].

When optimizing the right side of Eq. (36) by applying the genetic algorithm, M number of sample paths must be fixed. If the sample paths are not fixed, the evaluation value considerably fluctuates each time the objective function is evaluated, which would hinder optimization from progressing. However, when the sample path is fixed, the optimal solution only applies to those particular sample paths, and the objective function value deteriorates under other sample paths. Therefore, the following method is employed in this study. First, generate 100 bundles of M number of sample paths (i.e., $100M$ number of sample paths are generated). Next, apply the genetic algorithm under each bundle to obtain 100 optimal solutions. Lastly, obtain the general solution by taking the median value of these 100 optimal solutions. The median value is used instead of the average to reduce the influence of extreme values that appear because of the dispersion of random numbers.

The real-valued genetic algorithm requires the upper limit and lower limit of the variable $\tilde{s}_i$ $(1 \le i \le N_I)$. In the application of Section 5, the lower limit is set at 0°C, and the upper limit is set at 7.5°C. The number of sample paths in one bundle is set at $M = 1000$. Hence, the analysis is conducted with 100,000 sample paths.

5. Application

5.1 Setting of parameters

Herein, the proposed ROA model and the solution method were applied to the basin of an anonymous river in Japan. The conditions for the application are set as follows. The social discount rate ρ was set at 4%. The upper limit of the yearly budget $\overline{Y}$ was set at 5 billion yen. The parameters θ_{base} and η_{base} of the Gumbel distribution in Eqs. (9) and (10) were estimated using the maximum likelihood method to the observation of the yearly maximum rainfall from the past 50 years at the upstream of the target river. The value of u in Eqs. (9) and (10) was set to 0.05. This setting was based on the estimation by the Japanese Government [12], which evaluated the increase in the annual maximum rainfall in Japan by 1.1–1.15 times in RCP 2.6 scenario and 1.2–1.4 times in RCP 8.5 scenario.

The observed hyetographs of the past nine heavy rainfall events were used to construct the set of hyetographs $\{1, \cdots, N_W\}$ where $N_W = 9$. Nine residential areas

Residential area j	Economic damage [Billion yen]
1	410
2	2200
3	2600
4	1500
5	730
6	1300
7	570
8	300
9	630

Table 1.
Economic damage at each area when the worst-case flood scenario occurs.

were considered ($N_J = 9$). **Table 1** organizes the economic damage incurred if there is a rainfall for return period of 10,000 years in the current climate condition, the hyetograph with the sharpest peak is realized, and a levee breach occurs at each area j. The economic damage describes the damage to physical assets.

Eleven flood control facilities, namely one dam, two reservoirs (reservoir A and reservoir B), and the riverbed deepening in eight sections are considered as facilities yet to be constructed ($N_J = 11$). **Table 2** lists the parameters related to each facility and the residential areas on which each facility has an effect on risk reduction. The dam and the reservoirs will not have any effect of risk reduction until their construction has been completed, while the riverbed deepening will gradually show effect as construction advances. The upper limit of yearly construction cost $\bar{y}_i$ was set at 2.5 billion yen for every facility. While the dam and the reservoirs have a risk reduction effect on every residential area, the deepening of each river channel only has an effect on specific residential areas.

Facility i	Type of $f_i(x_i)$	c_i [bn. yen]	$\bar{y}_i$ [bn. yen]	Protected areas j
Dam	I_1	30	2.5	1–9
Reservoir A	I_1	10	2.5	1–9
Reservoir B	I_1	4	2.5	1–9
Deepening Section 1	I_2	2	2.5	1
Deepening Section 2–3	I_2	6	2.5	2, 3
Deepening Section 4	I_2	6	2.5	4
Deepening Section 5	I_2	2	2.5	5
Deepening Section 6	I_2	5	2.5	6
Deepening Section 7	I_2	2	2.5	7
Deepening Section 8	I_2	20	2.5	8
Deepening Section 9	I_2	9	2.5	9

Table 2.
Setting of flood control facilities yet to be constructed.

5.2 Effectiveness of the genetic algorithm

Figure 6 shows the conversion process of the genetic algorithm under a fixed sample path bundle. The horizontal axis represents the generation number (iteration number) of the genetic algorithm and the vertical axis demonstrates the objection function value. By the 51st generation, the objective function value has largely converged. The result confirms that the genetic algorithm functions appropriately.

5.3 Result of the analysis

Table 3 depicts the optimal solution $\tilde{s}_i$ obtained through the heuristic solution method proposed in Section 4.3. As the temperature change s_0 at $t = 0$ is 0.63 degrees,

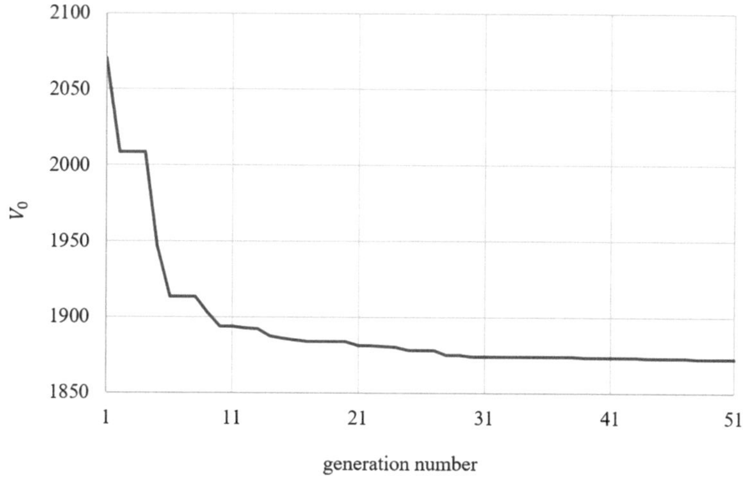

Figure 6.
Conversion process of the genetic algorithm.

Facility i	$\tilde{s}_i$
Dam	3.53
Reservoir A	3.05
Reservoir B	0.88
Deepening Section 1	0.30
Deepening Section 2–3	0.43
Deepening section 4	0.98
Deepening section 5	0.41
Deepening section 6	0.31
Deepening section 7	0.64
Deepening section 8	5.51
Deepening section 9	1.23
V_0 [billion yen]	174.08

Table 3.
Optimal solution $\tilde{s}_i$ and objective function value.

$\tilde{s}_i \leq 0.63$ indicates that the development of flood control facility i should be started now or in the near future. As defined by Eq. (31), flood control facility i with low $\tilde{s}_i$ have a high priority for development. At $t = 0$, the budget of 5 billion yen will be allocated to the flood control facilities for the riverbed deepening of Section 1 (1.5 billion yen), the riverbed deepening of Section 5 (1 billion yen), and the riverbed deepening of Section 6 (2.5 billion yen).

The optimal development period for flood control facilities is dependent on the temperature change that will be realized in the future. We specifically demonstrate this characteristic with two sample paths. The first is the sample path of the median temperature change between the RCP2.6 and the RCP8.5 scenarios shown in **Figure 1**. We call this the average scenario. The second is the sample path of the RCP8.5 scenario, as shown in **Figure 1**. **Figures** 7 and **8** show the period wherein each flood control facility is to be developed under these two sample paths. Each line of the table shows flood control facility i, and each column presents period t. The columns in the table are arranged in ascending order of time, with the leftmost column representing 2020 and the rightmost column representing 2100. The filled frame in each line indicates the period wherein that flood control facility is to be developed (the periods when $y_{ti} > 0$). Under the RCP8.5 scenario, the development of reservoir B and riverbed deepening at sections 4 and 9 are conducted earlier in contrast to that under the average scenario. Moreover, while the dam and reservoir A are not developed at all under the average scenario, they are developed after 2075 under the RCP8.5 scenario.

In this way, by utilizing the ROA model proposed in this study, it is possible to obtain a gradual and flexible development plan for flood control facilities that responds to the actual climate change monitoring situation. In the actual process of designing a flexible plan, we may have to think "it is better not to invest in this flood

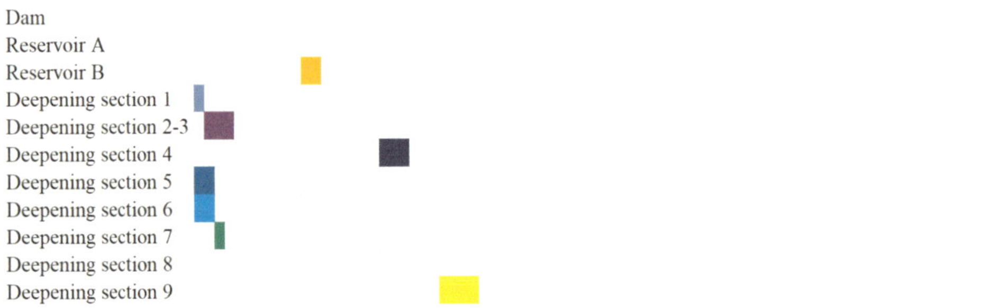

Figure 7.
Optimal development period of each facility under the average scenario.

Figure 8.
Optimal development period of each facility under the RCP8.5 scenario.

control facility because a large increase in temperature is unlikely once this situation realizes" or "it will be too late to start the construction of this flood control facility once the temperature has increased to this point," and so forth. The ROA model effectively enables the designing of development plans based on mathematical optimization.

Finally, we introduce the economic benefit of the flexible policy derived from the optimization problem. The objective function value when the optimal solution is adopted is 174.08 billion yen as shown in **Table 3**. Suppose the situation where we use inflexible planning that requires us to stick to the originally determined schedule to develop facilities, irrespective of the realization of climate change scenario. In this situation, the objective function value is 177.34 billion yen even if the development period is optimally determined. Therefore, the economic value of the flexible plan can be estimated at 3.26 billion yen. As this value shows, in situations where the future scenario is highly uncertain, allowing flexible planning has high economic value.

6. Conclusion

This study proposed an ROA model that dynamically optimizes budget allocation to flood control facilities while considering the uncertainty of climate change scenarios. The proposed model is applicable to actual river basins. Additionally, a heuristic solution method to this problem that utilizes a genetic algorithm is proposed. Moreover, it verified the effectiveness of the proposed model and solution method by applying them to an anonymous river basin in Japan. The result confirms that the flexible plan derived from the ROA model significantly reduces the social cost compared to inflexible planning.

Two challenges remain for this study. The first challenge is to relax the limitations mentioned in Section 3.7. The second challenge is to express the policy for determining the decision variable in a more complex functional form, for instance, through a neural network that uses all state variables as input.

Acknowledgements

This study was supported by Research and Development Grant No.19004 from the Japan Institute of Country-ology and Engineering, JSPS KAKENHI Grant Numbers JP 21K14265, JP23H00215, JP24K07700. We would like to express our gratitude for the grant.

Conflict of interest

The authors declare no conflict of interest.

Author details

Shunsuke Segi[1*], Katsumi Wakigawa[2], Toshiyuki Nishikori[2] and Yusuke Kobayashi[3]

1 Kobe University, Kobe, Japan

2 TOKEN C.E.E. Consultants Co., Ltd., Toshima-ku, Tokyo, Japan

3 Qammer Co., Ltd., Nihombashimuromachi, Chuo-ku, Tokyo, Japan

*Address all correspondence to: segi@people.kobe-u.ac.jp

References

[1] Rojas R, Feyen L, Bianchi A, Dosio A. Assessment of future flood hazard in Europe using a large ensemble of bias-corrected regional climate simulations. Journal of Geophysical Research Atmospheres. 2012;**117**:D17109. DOI: 10.1029/2012JD017461

[2] Alfieri L, Burek P, Feyen L, Forzieri G. Global warming increases the frequency of river floods in Europe. Hydrology and Earth System Sciences. 2015;**19**:2247-2260. DOI: 10.5194/hess-19-2247-2015

[3] Guerreiro SB, Dawson RJ, Kilsby C, Lewis E, Ford A. Future heat-waves, droughts and floods in 571 European cities. Environment Research Letters. 2018;**13**:034009. DOI: 10.1088/1748-9326/aaaad3

[4] Rojas R, Feyen L, Watkiss P. Climate change and river floods in the European Union: Socio-economic consequences and the costs and benefits of adaptation. Global Environmental Change. 2013;**23**:1737-1751. DOI: 10.1016/j.gloenvcha.2013.08.006

[5] Kind JM. Economically efficient flood protection standards for the Netherlands. Journal of Flood Risk Management. 2014;**7**:103-117. DOI: 10.1111/jfr3.12026

[6] Eijgenraam C, Kind J, Bak C, Brekelmans R, den Hertog D, Duits M, et al. Economically efficient standards to protect the Netherlands against flooding. Interfaces. 2014;**44**:7-21. DOI: 10.1287/inte.2013.0721

[7] Woodward M, Kapelan Z, Gouldby B. Adaptive flood risk management under climate change uncertainty using real options and optimization. Risk Analysis. 2014;**34**:75-92. DOI: 10.1111/risa.12088

[8] Kind JM, Baayen JH, Wouter Botzen WJ. Benefits and limitations of real options analysis for the practice of river flood risk management. Water Resources Research. 2018;**54**:3018-3036. DOI: 10.1002/2017WR022402

[9] Collins M et al. Long-term climate change: Projections, commitments and irreversibility. In: Stocker TF et al., editors. Climate Change 2013: The Physical Science Basis. Contribution of Working Group I to the Fifth Assessment Report of the Intergovernmental Panel on Climate Change. Cambridge and New York: Cambridge University Press; 2013. pp. 1029-1136

[10] Gosavi A. Simulation-Based Optimization – Parametric Optimization Techniques and Reinforcement Learning. 2nd ed. Springer; 2015. DOI: 10.1007/978-1-4899-7491-4

[11] Yoon Y, Kim Y-H. The roles of crossover and mutation in real-coded genetic algorithms. In: Gao S, editor. Bio-Inspired Computational Algorithms and their Applications. London, UK, London, UK: IntechOpen; 2012. pp. 65-82

[12] Ministry of Land, Infrastructure, Transport and Tourism: On Impact of Climate Change [Internet]. 2019 Available from: https://www.mlit.go.jp/river/shinngikai_blog/shaseishin/kasenbunkakai/shouiinkai/kikouhendou_suigai/1/pdf/09_kikouhendounoeikyou.pdf [Accessed: March 28, 2024]

Chapter 5

Comparative Analysis of Drought Indices for Soil Moisture Estimation in Nigeria's Diverse Climate Zones

Blessing Funmbi Sasanya, Akeem Abiodun Ademola and Oluwafemi Oladipupo Kolajo

Abstract

The understanding of soil moisture dynamics is important for agricultural productivity as well as water resource management. This study investigates the spatial variations of drought in relation to soil moisture, across three different soil layer depths in Nigeria, using the established drought indices. Multiple soil moistures at different levels within the soil profile were considered to investigate their respective changes between 1981 and 2022. Pearson's correlation coefficient (r) was used to assess the strength of complex relationship between the drought indices and historical soil moisture. Rainfall Anomaly Index (RAI) correlated best with soil moisture at 0–5 cm in 28 states, with r ranging from 0.52 to 0.86. Soil moisture in the arid areas correlated more closely with RAI but relates better with PHDI in the humid regions. This study would give invaluable insights to stakeholders for the enhancement of resilience in agricultural productivity and drought readiness in Nigeria.

Keywords: drought indices, Palmer Drought Severity Index (PDSI), self-calibrating Palmer Drought Severity Index (ScPDSI), Palmer Hydrological Drought Index (PHDI), soil moisture, Standardized Precipitation Index (SPI), Standardized Precipitation Evaporation Index (SPEI), Rainfall Anomaly Index (RAI), water resource planning and management, Pearson correlation coefficient

1. Introduction

Several parts of the world are facing climatic variabilities posed by global warming [1]. Climatic variables including precipitation and temperature are the basic and most important determinants of climatic variations of a particular region [2]. The monitoring and analysis of drought is of critical importance because drought is not instantaneous in nature [3]. Drought monitoring events are done effectively using several previously developed indices and indicators [4, 5]. The widely used drought indicators are precipitation, evaporation, temperature, soil moisture, among others [6]. Indices such as the Palmer Drought Severity Index (PDSI), Self-Calibrating Palmer Drought Severity Index (ScPDSI), Palmer Hydrological Drought Index (PHDI), Standardized Precipitation Index (SPI), Rainfall Anomaly Index (RAI), Rainfall Decile Index (RDI) and Standardized Precipitation-Evapotranspiration Index (SPEI) are employed in conjunction with the aforementioned indicators for effective drought monitoring. Such indices and indicators are important for insights on the location, severity and the time of drought [7].

One of the important factors in hydrology and vegetation restoration is soil moisture content especially in the arid regions where there is minimal groundwater storage [8]. Soil moisture therefore plays a major role in water, carbon and energy cycle [9]. Drought extent and severity can easily be indicated by soil moisture at different levels [10]. Soil moisture has been identified as very good fluctuating factor for agricultural drought monitoring [11–15]. The observations of soil moisture data are a challenge and studies usually make use of the relationship between soil moisture drought and other important variables [16, 17]. Soil moisture data can be obtained from *in situ* stations on a small scale but these are usually not enough for spatiotemporal analysis [8, 18]. Different land uses, precipitation patterns, vegetation cover and topography are included in the factors that affect soil moisture [19]. Previous studies have, however, suggested that vegetation cover has a declining effect on soil moisture due to the interception of rainfall by plant canopy and transpiration from plant leaf surfaces [20].

Drought results from minimal rainfall over a period of time, hence causing soil moisture reduction [21–23]. A very important aspect of drought is agricultural drought and it is determined by insufficiency of soil moisture over time and also has an effect on crop production [24]. On the other hand, drought occurrences resulting in deficiencies in reservoir levels and stream flow, over some period of time, are referred to as hydrological drought [25]. Reduced plant water uptake results from soil moisture reduction [26], and the limitation of plant water uptake and evapotranspiration (ET) tends to result in rise in heat flux [27] and inadequate crop nutrient uptake [28]. Socio-economic activities of a nation can be destabilized by the adverse effect of drought [25].

Having considered the potential adverse effect of drought in all socio-economic and nutritional aspects of nations, the objective of this study is thus to establish the level of complex relationship between the different drought indices and historical soil moisture at varying soil layer depths of 0–5 cm (top soil) (SMTop), 0–100 cm (root zone) (SMRoot) and 0–bedrock (surface and subsurface soils) (SMBed). This was achieved by correlating estimated drought indices with historical soil moisture data over 42 years (1981–2022) across the 36 six states and the Federal Capital territory of Nigeria. The drought indices employed include two multiscale models (Standardized Precipitation Evaporation Index (SPEI) and Standardized Precipitation Index (SPI) with five different time scales (variant)) and three two-layer bucket models (Palmer Drought Severity Index (PDSI), self-calibrated Palmer Drought Severity Index (ScPDSI) and Palmer Hydrological Drought Index (PHDI)) and one anomaly index (Rainfall Anomaly Index). These would enhance the adequate monitoring, and give proper insight and understanding of soil moisture variations over time. This would further enhance agricultural productivity alongside water resource management and optimization.

2. Materials and methods

2.1 Study area

Nigeria is one of the sub-Saharan African countries precisely located in West Africa between latitudes 4°16′N and 13°53′N, and longitudes 2°49′E and 14°37′E

(**Figure 1**) and bordered by countries including Benin republic, Chad, Cameroon and Niger. Nigeria occupies an approximate total land area of 923,768 km^2 and included within the country is coastline that stretches for 853 km along the Gulf of Guinea. Nigeria enjoys the tropical climate in the southern parts with two main seasons: the wet season that spans between April and October in most regions within the country and the dry season spanning from November to March [29]. The main climatic zones within the country are the tropical rainforest climate, the tropical savannah climate and the Sahel climate [29]. Nigeria has a mean annual and range temperature of 26.9° C and 24–30°C, respectively [30]. The temperature varies across the country, with coastal areas having ranges between 22°C and 32°C and inland areas between 18°C and 30°C. The southern and northern parts of Nigeria record their highest temperatures during the dry season, with the mean maximum temperatures of below 35°C in the south and above 40°C in the north [31].

2.2 Data source

Daily and monthly meteorological and soil moisture data spanning 42 years (1981–2022) were sourced from the official website of the National Aeronautics and Space Administration (NASA) for Nigeria as a whole. These data were provided in a global grid with a spatial resolution of 0.1° × 0.1° in latitude and longitude, equivalent to a grid size of approximately 10 kilometers [32]. Notably, previous studies by Refs. [32–34] have also utilized data from this source. The geographical coordinates of data stations in each state were determined using their administrative headquarters as reference points. The coordinates and elevation of all geographical locations where data are obtained are shown in **Table 1**. The sourced data included precipitation, minimum and maximum temperatures, wind speed and soil moisture at depths of 0–5 cm (top soil) (SMTop), 0–100 cm (root zone) (SMRoot) and 0–bedrock (subsoil)

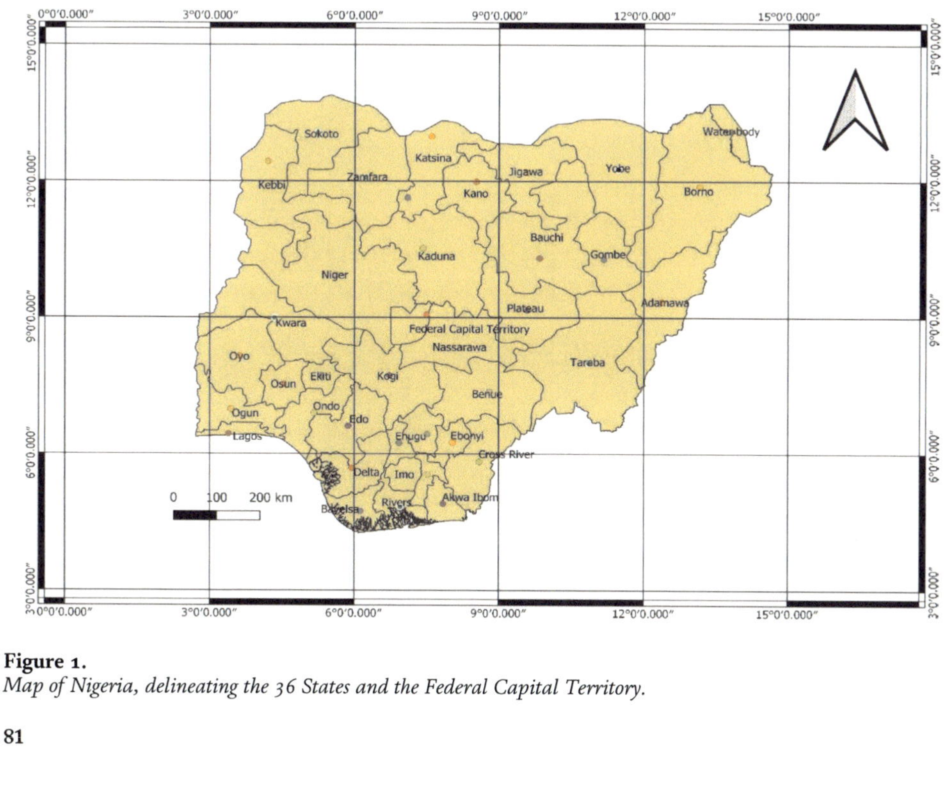

Figure 1.
Map of Nigeria, delineating the 36 States and the Federal Capital Territory.

S/N	State	Latitude (°N)	Longitude (°E)	Elevation ASL (m)	Average daily minimum temperature (°C)	Average daily maximum temperature (°C)	Average daily relative humidity (%)	Average daily precipitation (mm/day)
1	Abia	5.53	7.49	135.33	22.20	29.06	85.89	3.95
2	Abuja	9.06	7.50	491.64	20.46	30.62	71.54	5.30
3	Adamawa	9.26	12.45	283.16	21.66	33.90	56.18	2.66
4	Akwa-Ibom	5.03	7.93	37.19	23.16	28.86	86.92	8.09
5	Anambra	6.22	7.08	84.12	22.20	29.12	84.89	5.24
6	Bauchi	10.31	9.85	608.08	19.96	33.17	50.73	2.52
7	Bayelsa	4.76	6.12	9.75	23.08	29.05	87.62	7.41
8	Benue	7.74	8.52	128.63	21.61	30.27	77.28	3.80
9	Borno	11.84	13.14	309.68	21.36	35.77	39.34	1.55
10	Cross River	4.95	8.31	187.15	23.16	28.86	86.92	8.09
11	Delta	6.17	6.73	17.07	22.20	29.12	84.89	5.25
12	Ebonyi	6.33	8.11	211.53	21.94	29.68	82.04	4.30
13	Edo	6.63	5.92	234.70	22.18	29.15	86.93	5.18
14	Ekiti	7.72	5.31	436.78	20.90	28.91	83.11	4.56
15	Enugu	6.44	7.49	209.40	21.67	29.30	82.60	4.49
16	Gombe	10.29	11.16	354.79	20.55	33.24	53.82	2.76
17	Imo	5.57	7.04	133.50	22.46	29.33	85.97	5.70
18	Jigawa	12.23	9.58	373.08	19.79	34.78	43.25	1.97
19	Kaduna	10.53	7.43	613.56	18.57	30.10	65.52	4.80
20	Kano	11.99	8.52	449.88	19.21	34.43	45.53	2.04
21	Katsina	12.99	7.60	519.68	20.14	33.81	41.20	1.88

DOI: http://dx.doi.org/10.5772/intechopen.1007115

S/N	State	Latitude (°N)	Longitude (°E)	Elevation ASL (m)	Average daily minimum temperature (°C)	Average daily maximum temperature (°C)	Average daily relative humidity (%)	Average daily precipitation (mm/day)
22	Kebbi	13.25	4.30	195.99	21.74	35.57	40.09	1.54
23	Kogi	7.68	6.75	167.94	31.81	30.56	77.35	3.84
24	Kwara	8.99	4.35	310.29	21.47	32.38	68.58	2.82
25	Lagos	6.45	3.39	5.79	24.46	28.63	84.60	4.05
26	Nasarawa	11.64	7.10	638.86	21.47	31.06	72.54	3.82
27	Niger	9.82	5.64	39.62	20.49	31.46	69.58	3.62
28	Ogun	6.99	3.55	135.64	22.52	29.34	86.42	4.60
29	Ondo	6.89	5.18	176.17	21.71	28.83	87.05	4.94
30	Osun	7.54	4.54	287.12	21.45	29.28	84.53	4.78
31	Oyo	8.14	3.62	348.69	21.17	29.68	81.41	4.28
32	Plateau	9.20	9.57	387.10	20.70	31.16	62.22	3.13
33	Rivers	4.84	6.95	13.41	22.64	29.37	87.43	7.23
34	Sokoto	13.06	5.24	287.73	21.30	34.91	43.69	1.82
35	Taraba	8.00	10.84	178.92	21.58	31.24	71.99	4.47
36	Yobe	12.26	11.46	357.23	20.92	35.20	37.89	1.55
37	Zamfara	12.15	6.23	399.29	20.31	33.50	51.35	2.35

Table 1.
Coordinates, elevation and meteorological data of study areas.

(SMBed). The daily temperature and wind speed data were employed to estimate daily evapotranspiration, using the FAO penman Monteith Equation, for subsequent aggregation into monthly evapotranspiration data. The monthly precipitation and evapotranspiration data were, however, used as input for the estimation of drought indices.

2.3 Drought indices

2.3.1 Palmer Drought Severity Indices (PDSI) and self-calibrated Palmer Drought Severity Indices (ScPDSI)

The PDSI and ScPDSI are simple two-layer bucket models that simulate the balance between moisture supply (precipitation) and moisture demand (evapotranspiration) in the soil and atmosphere [35]. The Palmer Drought Severity Index was from estimated from all data stations in the 36 states and Abuja in Nigeria. The estimation of PDSI is dependent on Eq. (1).

$$p = \alpha PET + \beta PR + \gamma PRO – \delta PL \tag{1}$$

Where "p" is the amount of rain needed to keep the soil moist, "PET" is the amount of water that evaporates from the soil and plants, "PR" is the amount of water that recharges the soil, and "PRO" is the amount of water that runs off the soil surface. "PL" represents the water lost, and "α", "β", "γ" and "δ" are special factors that enable the estimation of these parameters. These special factors were estimated from Eq. (2).

$$\alpha = \frac{ET}{PET}; \beta = \frac{R}{PR}; \gamma = \frac{RO}{PRO}; \delta = \frac{L}{PL} \tag{2}$$

$$d = P - p = P - (\alpha PE + \beta PR + \gamma PRO - \delta PL) \tag{3}$$

Moisture departure (d), which is the difference between the real amount of rainfall (P) in a particular month and the predicted amount (p) under the prevailing circumstances, was calculated from Eq. (3). To ensure a strong correlation, the moisture deviation was adjusted by a climate characteristic coefficient (K) (Eq. (4)). The product of K and d resulted in the moisture anomaly index, Z_i. This index reflects the level of wetness or dryness within a specific month, regardless of the precipitation patterns [36]. From each Z index for a given month, PDSI value (X_i) was derived from Eq. (5).

$$K = \frac{17.67}{\sum_{j-1}^{12} d_j K_j} K_j \tag{4}$$

$$X_i = 0.897 X_{i-1} + \left(\frac{1}{3}\right) Z_i \tag{5}$$

Xi denotes the PDSI value of the current month, while X(i − 1) represents the PDSI value of the previous month, and the constants 0.897 and 1/3 are the duration factors. **Table 2** depicts the classes and grades for the classification of PDSI, SPEI and SPI values.

The Self-Calibrating Palmer Drought Severity Indices (ScPDSIs) was calculated from similar sets of equations; however, the Z index used for ScPDSI was modified

PDSI/ScPSDI/PHDI	Wet and dry grade	SPEI/SPI	RAI	Class description
≥4	Extremely wet	≥ 2	≥3	Extremely wet
≥3 ∼ 4	Very wet	1.5–1.99	2–2.99	Very wet
≥2 ∼ 3	Moderately wet	1–1.49	0.5–1.99	Moderately wet
≥1 ∼ 2	Slightly wet	−0.99 to −0.9	−0.49 to −0.49	Normal
≥-2 ∼ -1	Mild drought	- 1 to −1.49	−1.99 to −0.5	Moderately dry
≥-3 ∼ -2	Moderate drought	−1.5 to −1.99	−2.99 to −2	Very dry
≥-4 ∼ -3	Severe drought	<= −2	<= −3	Extremely dry
<-4	Extreme drought			

Source: [37–39].

Table 2.
Drought indices (PDSI, PHDI, SPEI, SPI and RAI) dryness and wetness grades.

and adjusted for local climate peculiarities and variabilities in temperature and pressure based on their means and standard deviations also taken into account [40]. Both PDSI and ScPDSI were estimated from the ScPDSI package in r programming 4.0.

2.3.2 Palmer Hydrological Drought Indices (PHDI)

The Palmer Hydrological Drought Index (PHDI) measures the severity of hydrological drought, which is the lack of water resources in an environment [41]. The PHDI is considered a long-term measure of hydrological drought because indicators such as river flows, reservoir water levels and groundwater levels often stay below average for a while after a meteorological drought has ended. In this study, the PHDI values were calculated from Palmer Drought Severity Index (PDSI) using Eq. (6).

$$PHDI = 0.897PDSI + 0.164PHDI_{t-1} \tag{6}$$

However, the PHDI takes into account changes in precipitation and runoff as well as the storage and movement of water in the soil, groundwater, and surface water bodies, making it a more comprehensive measure of drought that uses additional weighting factors during calculation. Importantly, the PHDI measures hydrological drought (water resource deficiency), whereas the PDSI measures meteorological drought (precipitation deficiency) [42]. Both PHDI and PDSI were estimated from the ScPDSI package in r programming 4.0.

2.3.3 Standardized Precipitation Evaporation Index (SPEI)

The Standardized Precipitation Evapotranspiration Index (SPEI) is dependent on evapotranspiration and precipitation data. SPEI has great advantage, because it incorporates various factors to assess drought, including the effects of temperature variability, wind speed and location, among others [43]. The SPEI values were obtained from the differences between the monthly potential evapotranspiration (PET) in mm and rainfall (mm). The PET was estimated using the FAO Penman Monteith model.

The estimation of the value of Di was conducted at intervals of 1 month, 3 months, 6 months, 9 months and 12 months for a period of 42 years. The difference for a particular month j and year i is influenced by the chosen time steps (k), which in this case ranges between 1 and 12. According to Ogunrinde et al. [36], the equations for the time steps are provided as Eqs. (7) and (8).

$$X_{ij}^{k} = \sum_{i=13-j-k}^{12} D_{i-1,l} + \sum_{i=1}^{j} D_{i,l} \qquad \text{for } j \leq k \text{ and} \tag{7}$$

$$X_{ij}^{k} = \sum_{i=j-k+1}^{j} D_{i-1,l} \qquad \text{for } j \geq k \tag{8}$$

The difference (D_i) was thereafter integrated to a three-factor log-logistic distribution. The probability density function of the three-factor log-logistic distribution is given as Eq. (9) and the cumulative density function of the three-factor log-logistic distribution is given as Eq. (10):

$$f(x) = \frac{\beta}{\alpha}\left(\frac{x-\gamma}{\alpha}\right)^{\beta-1}\left(1+\left(\frac{x-\gamma}{\alpha}\right)^{\beta}\right)^{-2} \tag{9}$$

$$F(x) = \left(1+\left(\frac{x-\gamma}{\alpha}\right)^{\beta}\right)^{-1} \tag{10}$$

where α is the scale parameter, β represents shape parameter and γ represents location parameter of the log-logistic distribution for values of D with range $\gamma > D > \infty$. The maximum-likelihood estimate method was used to determine the parameters of the log-logistic distribution. **Table 1** details the classes and grades for the classification of SPEI values.

2.3.4 Standardized Precipitation Index (SPI)

The Standardized Precipitation Index (SPI) was employed to gauge the level of wetness and dryness in the study area. SPI serves as a valuable tool for evaluating drought conditions over extended durations [44]. This involves subjecting the data to probability distribution through the use of the gamma function, particularly for long-term precipitation assessments [45, 46]. This index has demonstrated its ability to provide a dependable estimate of drought severity for spatial analyses [47]. Normalized numerical values are assigned to long-term data, with the deviation from the long-term mean of the precipitation under examination for the random variables that follow a normal distribution being represented as the SPI [39, 48].

When precipitation exceeds the long-term average, the SPI becomes positive; otherwise, it is negative. SPI values for each time step (1, 3, 6, 9 and 12 months) were derived by fitting precipitation data to the Cumulative Density Function (CDF) of the gamma distribution [39, 49], given as Eqs. (11)–(15).

$$G(x) = \int_{0}^{x} \frac{1}{\beta^{\propto}\Gamma(\propto)} x^{\propto-1} e^{-\left(\frac{x}{\beta}\right)}\, dx \tag{11}$$

$$\propto = \frac{1}{4A}\left(1+\sqrt{1+\frac{4A}{3}}\right) \tag{12}$$

$$\beta = \frac{X_a}{\alpha} \tag{13}$$

$$A = \ln(X_a) - \frac{\ln(\sum(\overline{X}))}{n} \tag{14}$$

$$H\ (x) = q + (1 - q)\ G(x) \tag{15}$$

where n is the number of years, α and β are the respective scale and shape parameters of the CDF, x is the monthly precipitation amount (mm), and $\overline{X}$ is the average value of the annual precipitation (mm). The gamma distribution for a zero precipitation amount is not defined and is thus taken care of by Eq. (15), and q is the probability of zero value which is given as m/n, where m = number of zero values and n = total number of observations. The SPI values obtained are classified as shown in **Table 1**.

2.3.5 Rainfall Anomaly Index (RAI)

The Rainfall Anomaly Index (RAI), developed by Van Rooy in 1965, is a rank-based drought index that quantifies the magnitude of positive and negative precipitation anomalies using a ranking procedure [50]. This statistical method measures the deviation of rainfall from its long-term average by assessing its variability [51]. RAI is particularly effective in humid to moderate climates, where monthly precipitation is relatively evenly distributed throughout the year, and the associated rainfall distribution is less skewed [52]. The modified RAI (mRAI) was therefore proposed for use in arid regions where there are frequent occurrences of zero monthly precipitation [52]. However, in this study, the RAI was employed and calculated using Eqs. (16) and (17).

$$RAI = +3\left(\frac{Ni - \overline{N}}{E_1 - \overline{N}}\right) \tag{16}$$

$$RAI = -3\left(\frac{Ni - \overline{N}}{E_2 - \overline{N}}\right) \tag{17}$$

Where: N_i denotes the measured precipitation, $\overline{N}$ represents the average precipitation, E_1 and E_2 stand for the average of 10 extremes and +3 indicates the prefix utilized to define the upper bounds of anomalies, while −3 defines the lower bound of the anomalies.

To estimate E_1 and E_2, the 504 monthly rainfall data points were arranged in descending order of magnitude. E_1 was calculated as the average of the 10 highest values, while E_2 was determined as the average of the 10 lowest values. Eq. (16) was then used to calculate the RAI for monthly rainfall amounts that exceed the mean value ($\overline{N}$) and Eq. (17) was applied to estimate the RAI for rainfall amounts that fall below the mean.

2.4 Pearson correlation coefficient

The Pearson correlation analysis assumes the data being analyzed have linear relationships, and are independent, normally distributed and homoscedastic [53]. Prior to subjecting the data set to the Pearson correlation test, the data were tested for normality using the Shapiro-Wilk's test. The Pearson correlation was performed at 5%

level of significance. In other words, when the p value was less than or equal to 0.05, it indicated that the level of correlation between the compared variables is significant.

The Pearson correlation coefficient (r) as a statistical technique was used to quantify the strength and direction of the linear relationships and the degree of association between each drought indices and the soil moisture at various depths of 0–5 cm (SMTop), 0–100 cm (SMRoot) and 0–bedrock (SMBed). The Pearson correlation coefficient (r) was estimated from Eq. (18) and it ranges from −1 to 1, where 1 signifies a perfect positive linear relationship, −1 depicts a perfect negative linear relationship and 0 indicates no linear relationship. The strength of correlation is determined by the magnitude of the value of r. The ranges 0–0.19, 0.20–0.39, 0.40–0.69 and 0.70 to 1.0 depict very week, weak, moderate and strong correlations, respectively [54].

$$r = \frac{\left[\frac{\Sigma(x_i-\bar{x})(y_i-\bar{y})}{(n-1)}\right]}{\sqrt{\left[\frac{\Sigma(xi-\bar{x})^2}{(n-1)}\right]\left[\frac{\Sigma(yi-\bar{y})^2}{(n-1)}\right]}} \tag{18}$$

Where r is the Pearson's correlation coefficient, x is the first variable, y is the second variable, $\bar{x}$ is the mean of the first variable, $\bar{y}$ is the average of the second variable and n is the number of data which in this case is 504, that is, monthly data for 42 consecutive years.

3. Results and discussion

3.1 Results

According to Sims et al. [55], the Standardized Precipitation Index (SPI) is better suited to represent short-term precipitation and soil moisture variations. In comparison with the Palmer Drought Severity Index (PDSI), SPI has been suggested as a viable means of estimating soil moisture in North Carolina. This study compares and correlates variants of different drought indices with soil moisture values at three different depths (0–5 cm [SMTop], 0–100 cm [SMRoot], and 0-bedrock [SMBed]) across the 36 states of Nigeria and the Federal Capital Territory (FCT). The drought indices include PDSI, ScPDSI, PHDI, variants time scales of SPI (1-month, 3-month, 6-month, 9-month and 12-month), variants time scales of SPEI (1-month, 3-month, 6-month, 9-month and 12-month) and Rainfall Anomaly Index (RAI). The maximum ranges of the drought indices across the study areas are 41.03, 12.00, 41.03, 8.18, 8.50, 8.30, 7.63, 7.53, 8.03, 7.03, 7.11, 7.76, 7.66 and 36.11 respectively. Conversely, the minimum ranges are 9.88, 8.83, 9.88, 5.75, 5.29, 5.04, 4.92, 4.40, 4.83, 4.52, 4.45, 4.09, 4.00 and 6.23, respectively. High values of PHDI and PDSI were recorded in dryer regions that experience with minimal rainfall and annual rainfall. Such areas include Kaduna (41.03), Katsina (28.83) and Sokoto (25.43). In contrast, states with high annual rainfall records contributed to the minimum PHDI and PDSI ranges. Examples of such states are Lagos (9.88), Rivers (13.64), Bayelsa (13.93) and Delta (12.31). **Figures 2–7** represent the mean values, standard error, maximum and minimum values of each drought index for 37 locations across the country over a period of 41 years (1981–2022). The mean PDSI, ScPDSI and PHDI across the entire nation ranged from −0.425 to 0.063 (**Figure 2**), −1.12 to −0.02 (**Figure 3**) and − 0.307 to

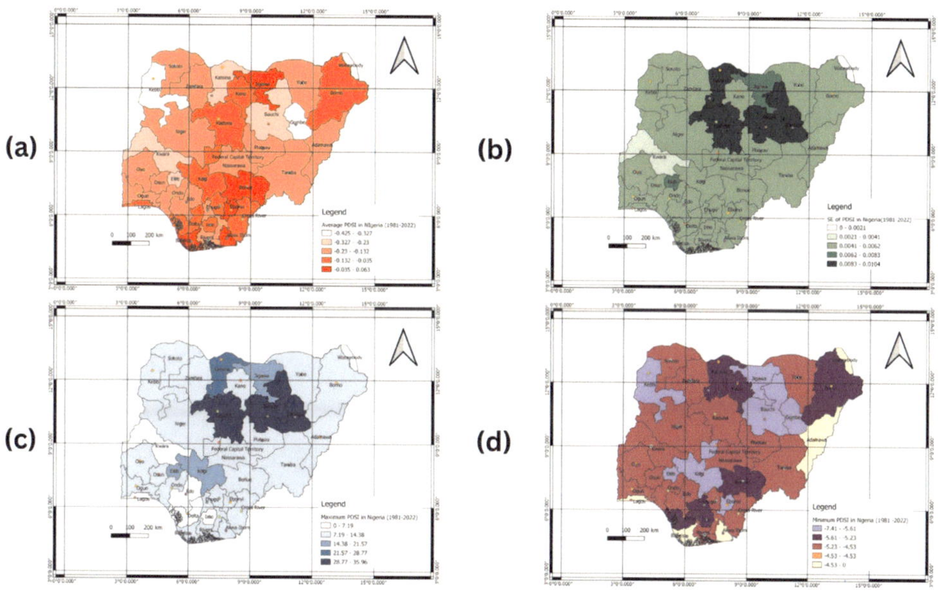

Figure 2.
Averages, standard errors, maximum and minimum values of Palmer Drought Severity Indices (PDSI) across Nigeria.

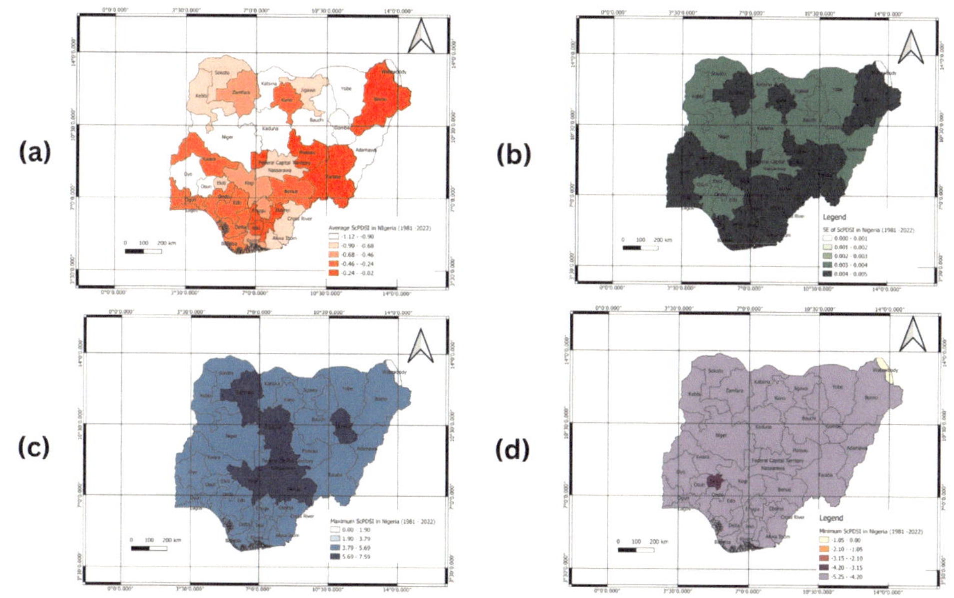

Figure 3.
Averages, standard errors, maximum and minimum values of self-calibrated Palmer Drought Severity Indices (ScPDSI) across Nigeria.

0.153 (**Figure 4**). These indicate that the entire Nigerian space, from the northern to southern regions, can be wholly classified as slightly wet to moderately dry.

Undoubtedly, Nigeria's geographical space encompasses areas with extremely dry, moderately dry, slightly wet and highly wet conditions. However, the average of these

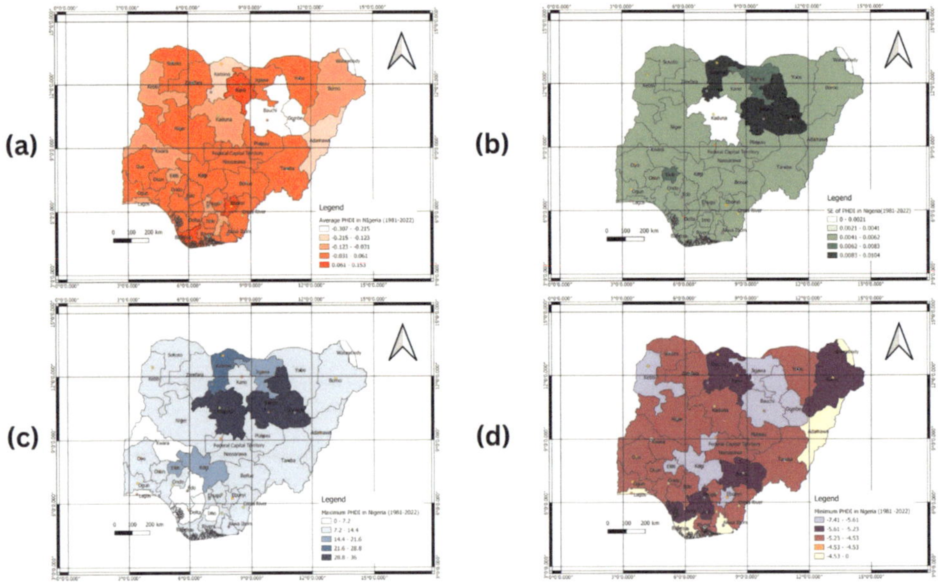

Figure 4.
Averages, standard errors, maximum and minimum values of Palmer Hydrological Drought Indices (PHDI) across Nigeria.

diversities results in a slightly wet and moderately dry environment. Furthermore, the averages of 1-month SPEI, 1-month SPI and RAI ranged from −0.008 to 0.999 (**Figure 5**), 0.001 to 0.241 (**Figure 6**) and −1.55 to −0.63 (**Figure 7**), respectively. These values support the notion that the average Nigeria's environment can be characterized as ranging between moderately dry, normal and moderately wet conditions. Notably, the low-average range of PDSI (−0.425 to −0.327) resulted from Kebbi and Gombe states (**Figure 2**). Similarly, the averages of PHDI for Bauchi and Gombe were the lowest (−0.307 to −0.215).

As shown in **Figure 5**, Kaduna, Ekiti and Akwa-Ibom exhibited similar SPEI values (0.014–0.035). In line with the obtained averages of PHDI for Kaduna and Gombe, both areas also had RAI values within the same range (−1.37 to −1.18). Kaduna, Bauchi and Gombe recorded the highest PHDI and PDSI values (28.77–35.96). **Figures 5** and **6** reveal that the maximum 1-month SPEI, 1-month SPI and RAI values are clustered. Specifically, the 1-month SPEI values obtained from the FCT, Kebbi, Kwara and Gombe states range from 0.91 to 1.82.

The standard error (SE), an indicator of the accuracy with which a population is represented by samples [56], compares the variability between different samples and indicates the precision level of sample estimates [57]. Smaller SE values indicate more precise estimates [58]. The standard errors of the means of selected drought indices are minimal, as shown in **Figures 2–7**. The SE of the means for PDSI and PHDI for most states ranges from 0.0041 to 0.0062. Similarly, the SE of means for 1_month SPEI and 1_month SPI ranges from 0.00159 to 0.00199, and the SE of means for RAI lies between 0.00307 and 0.00384. As depicted in **Figures 2–7**, the minimum PDSI, ScPDSI, PHDI, 1_month SPEI, 1_month SPI and RAI values are also clustered, with most states exhibiting PDSI, ScPDSI and PHDI values between −5.23 and −4.2. Similarly, the minimum 1_month SPEI values in most federal states are within the −3.17

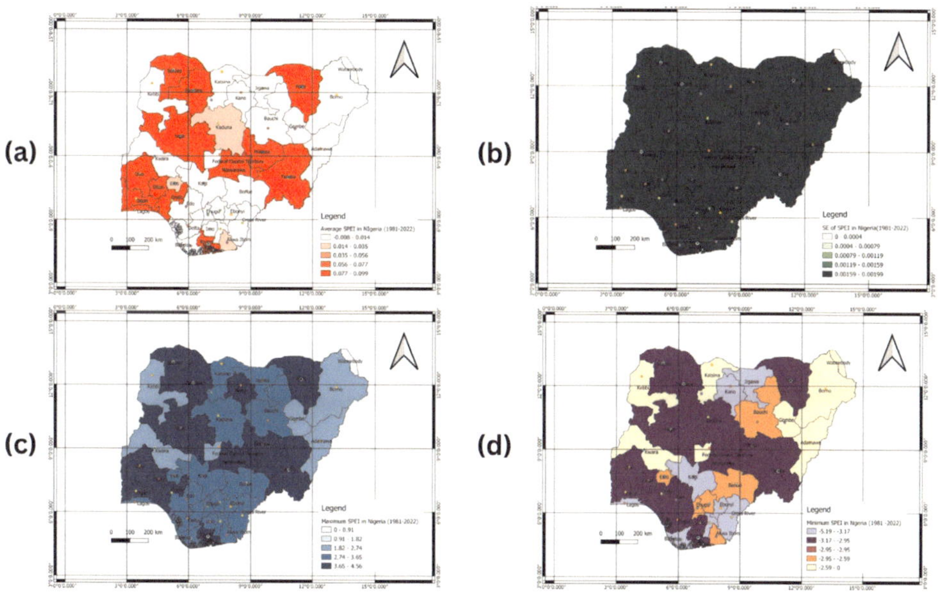

Figure 5.
Averages, standard errors, maximum and minimum values of 1 month Standard Precipitation Evaporation Indices (1_SPEI) across Nigeria.

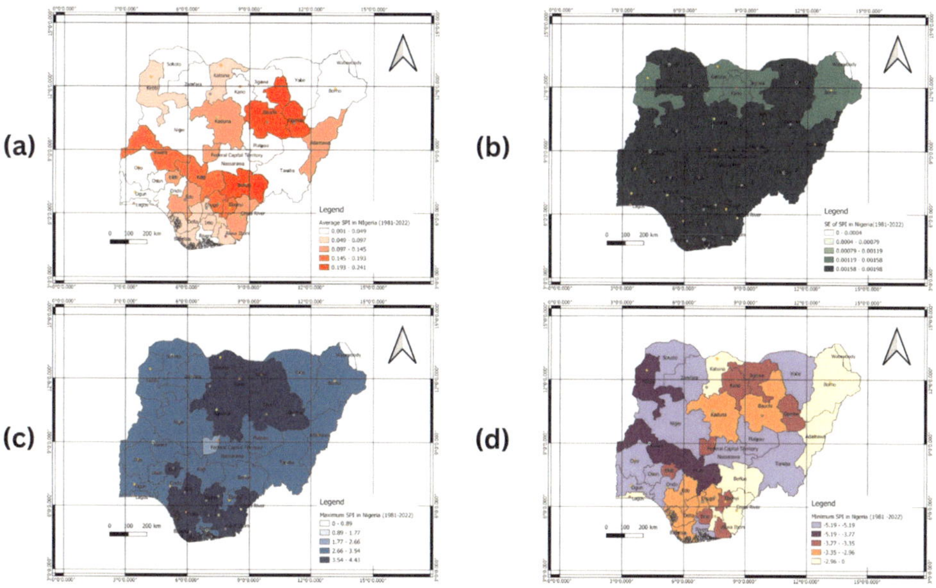

Figure 6.
Averages, standard errors, maximum and minimum values of 1 month Standard Precipitation Indices (1_SPI) across Nigeria.

and −2.95 boundaries, while the minimum 1_month SPI value for most states is −5.19, and the minimum RAI in all states ranges from −3.1 to −2.1.

Figures 8–10 illustrate the distribution of monthly averages of soil moisture across Nigeria. Specifically, **Figure 8** presents the mean, standard error (SE), maximum and

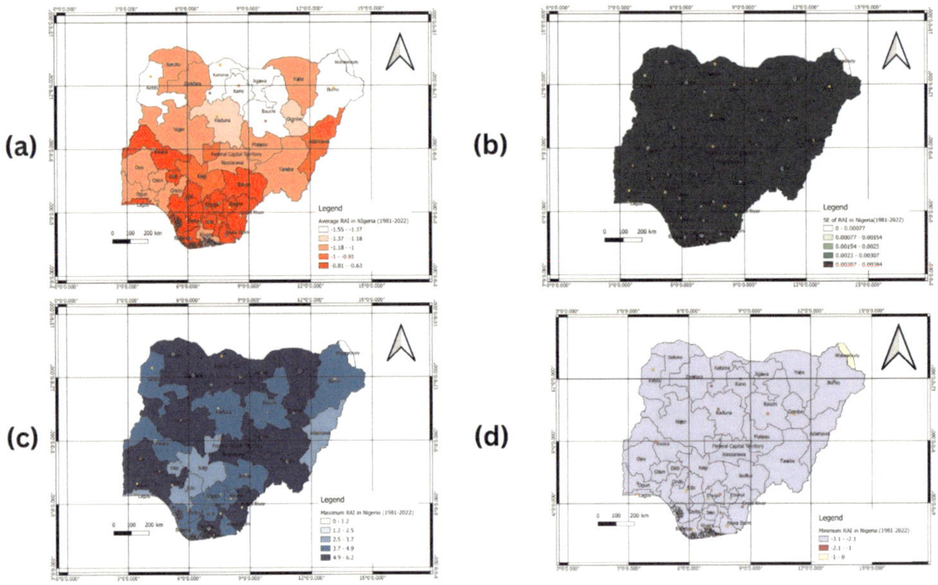

Figure 7.
Averages, standard errors, maximum and minimum values of Rainfall Anomaly Indices (RAI) across Nigeria.

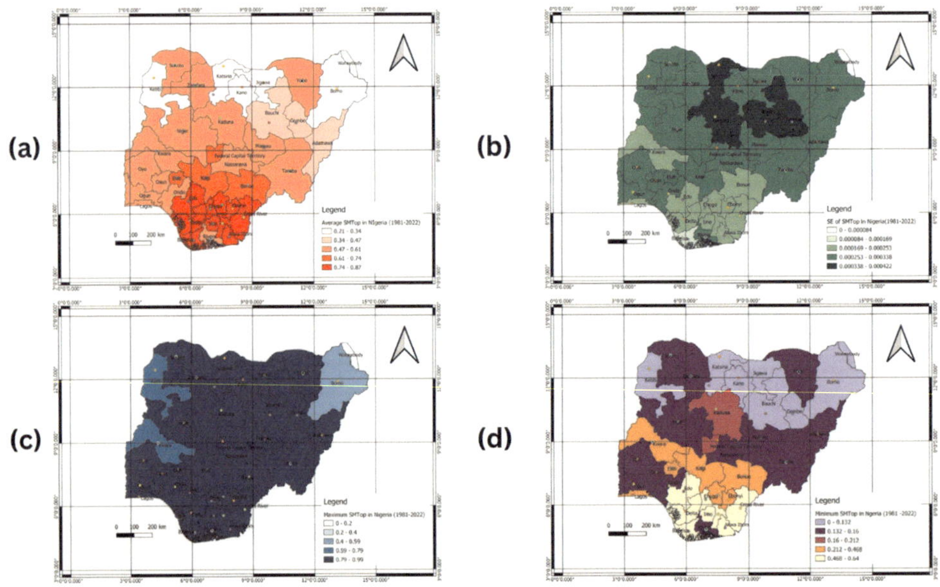

Figure 8.
Averages, standard errors, maximum and minimum values of top soil moisture (0–5 cm) (SMTop) across Nigeria.

minimum soil moisture at the first 5 cm depth (SMTop). **Figures 9** and **10** represent the mean, SE, maximum and minimum soil moisture at the first 100 cm or root zone (SMRoot) and from 0 cm to the bedrock (SMBed), respectively. The average SMTop values, as depicted in **Figure 8**, exhibit an increasing trend from the northern region (0.21–0.34) towards the southern region (0.74–0.87). This trend is also observed in

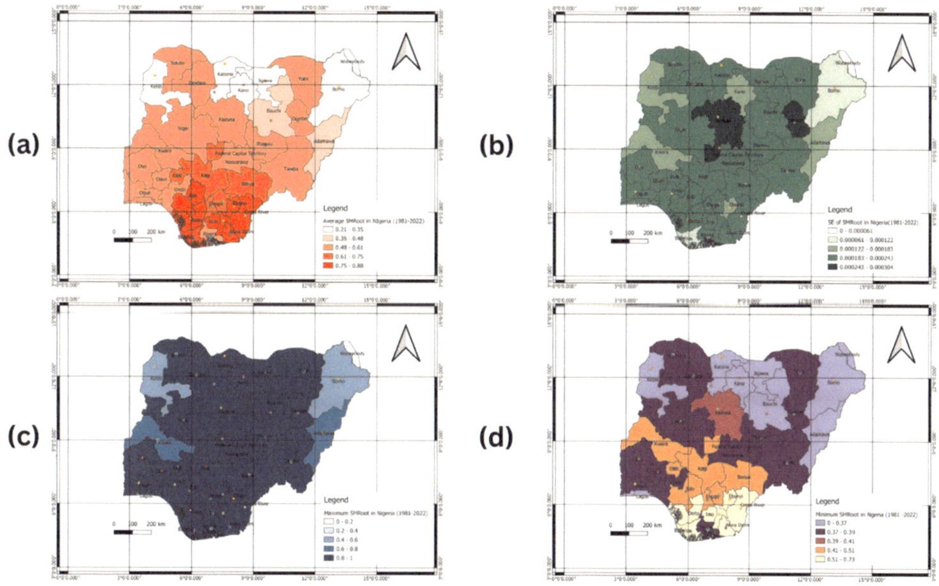

Figure 9.
Averages, standard errors, maximum and minimum values of root zone soil moisture (0–100 cm) (SMRoot) across Nigeria.

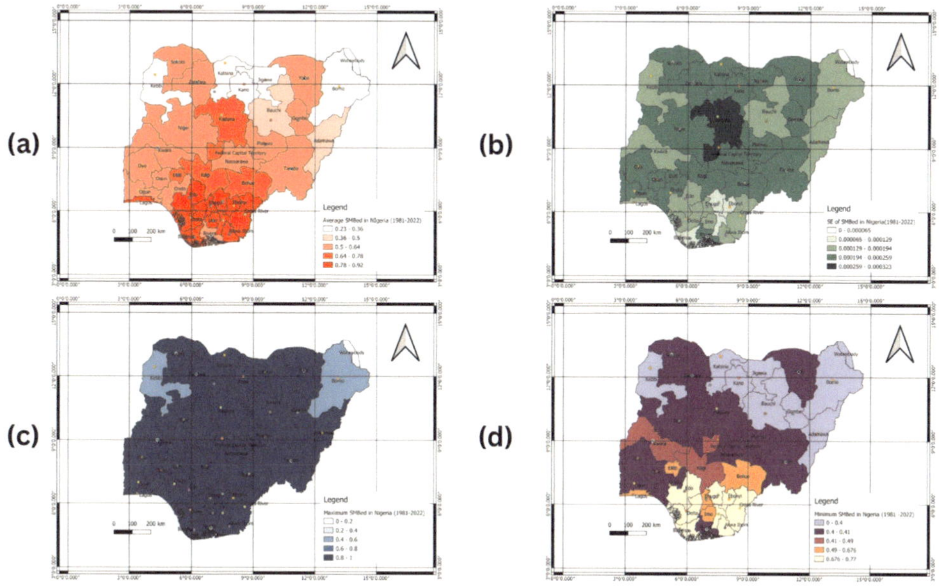

Figure 10.
Averages, standard errors, maximum and minimum values of surface and subsoil moisture (0 - Bedrock) (SMBed) across Nigeria.

the minimum SMTop values, with the far northern region having the lowest soil moisture storage (0–0.132) and the southern region having higher values of soil moisture storage (0.468–0.640) between soil layers 0 and 5 cm. Similar distribution trends were also evident in SMRoot and SMBed. The maximum SMTop values in most states range from 0.79 to 0.99 (**Figure 8**), while the maximum SMRoot and SMBed values range from 0.8 to 1 in over 90% of the states.

To investigate the most effective drought indices for estimating or predicting soil moisture level and probability of agricultural drought with the view of enhancing agricultural crop yield, variants of drought indices were correlated with soil moisture stored at different levels, as illustrated in **Figure 11**. Correlation analysis was conducted for each federal state, as shown in **Figure 10**. Observations from the figure revealed that PDSI, ScPDSI and PHDI, time lag variants of SPEI and SPI as well as SMTop, SMBed, and SMRoot exhibited significantly high correlations at $p < 0.05$ across all federal states among the groups. However, significantly low correlations were observed between RAI and the other four drought indices. Furthermore, in some states, the low correlation between RAI and the other drought indices was non-significant at $p < 0.05$. These findings suggest that most time lag variants of SPEI and SPI are not suitable for estimating soil moisture in most Nigerian states, whereas RAI, PHDI, PDSI and the 9 and 12 months' time scale variants of SPI may have more applicability in this context, depending on the prevalent climatic condition and the soil layer under consideration.

Figure 11 illustrates that the drought indices best correlate with stored soil moisture across each state and soil depths. The set of RAI values obtained for 16 states correlated best with the soil moisture stored at the three soil layers. In contrast, PDSI and PHDI values estimated for six states correlated best with soil moisture stored at the three different levels, while different indices had best correlation for different soil layers in 14 states and the FCT. RAI correlated best with SMTop in 28 states, with Pearson's correlation coefficient (r) ranging from 0.52 to 0.86. PHDI exhibited the strongest correlation with SMTop in 6 states, with r ranging from 0.53 to 0.70. PDSI only correlated best with SMTop in Edo state with $r = 0.66$, while 9_SPI correlated best with SMTop at Ekiti ($r = 0.63$) and Kogi ($r = 0.62$). Pearson's correlation coefficient of RAI with SMRoot was the strongest in 23 states, with r ranging from 0.51 to 0.77. PHDI correlated best with SMRoot in 12 states, with r ranging from 0.33 to 0.70. PDSI correlated best with SMRoot in Edo ($r = 0.65$) and Imo ($r = 0.53$) states.

The correlation coefficient between RAI and SMBed is the strongest in Abia (0.76), Adamawa (0.71), Akwa Ibom (0.74), Anambra (0.69), Bayelsa (0.73), Benue (0.70), Borno (0.64), Cross River (0.42), Delta (0.42), Gombe (0.55), Jigawa (0.54), Katsina (0.67), Kebbi (0.62), Plateau (0.66), Rivers (0.52), Taraba (0.51), Sokoto (0.50) and Yobe (0.62). In more than half of the states, SMBed exhibits the best correlation with PHDI, specifically in FCT (Abuja) (0.60), Bauchi (0.70), Cross River (0.51), Delta (0.51), Ebonyi (0.59), Edo (0.69), Enugu (0.56), Kano (0.54), Kogi (0.63), Kwara (0.56), Niger (0.56), Ogun (0.67), Ondo (0.57), Osun (0.47), Oyo (0.51) and Zamfara (0.51). Notably, PDSI exhibits the best linear correlation with SMBed in Imo state. Furthermore, Pearson's correlation coefficient of SMRoot with PHDI and PDSI has equal correlation coefficient (0.55) in Bauchi, and similarly, the correlation coefficient of SMRoot with PHDI and RAI has equal correlation coefficient (0.53) in Kano. ScPDSI correlated best with SMBed at Enugu, Kaduna and Lagos with $r = 0.56$, 0.52 and 0.69, respectively. In the same vein, 9_SPI had best correlation with SMBed at Ekiti, Nasarawa and Niger. These findings suggest that no single drought index is universally applicable for predicting or estimating soil moisture across Nigeria at different soil layer depths. Instead, RAI correlates best with SMTop and SMRoot in most locations, while PHDI correlates best with SMBed in more than half of the locations considered. As the soil layer depth increases, RAI becomes less applicable, while the applicability of double-layer bucket models increases (**Figure 11**). SPEI and SPI were observed to have exhibited better correlation with soil moisture as the time lag increases. However,

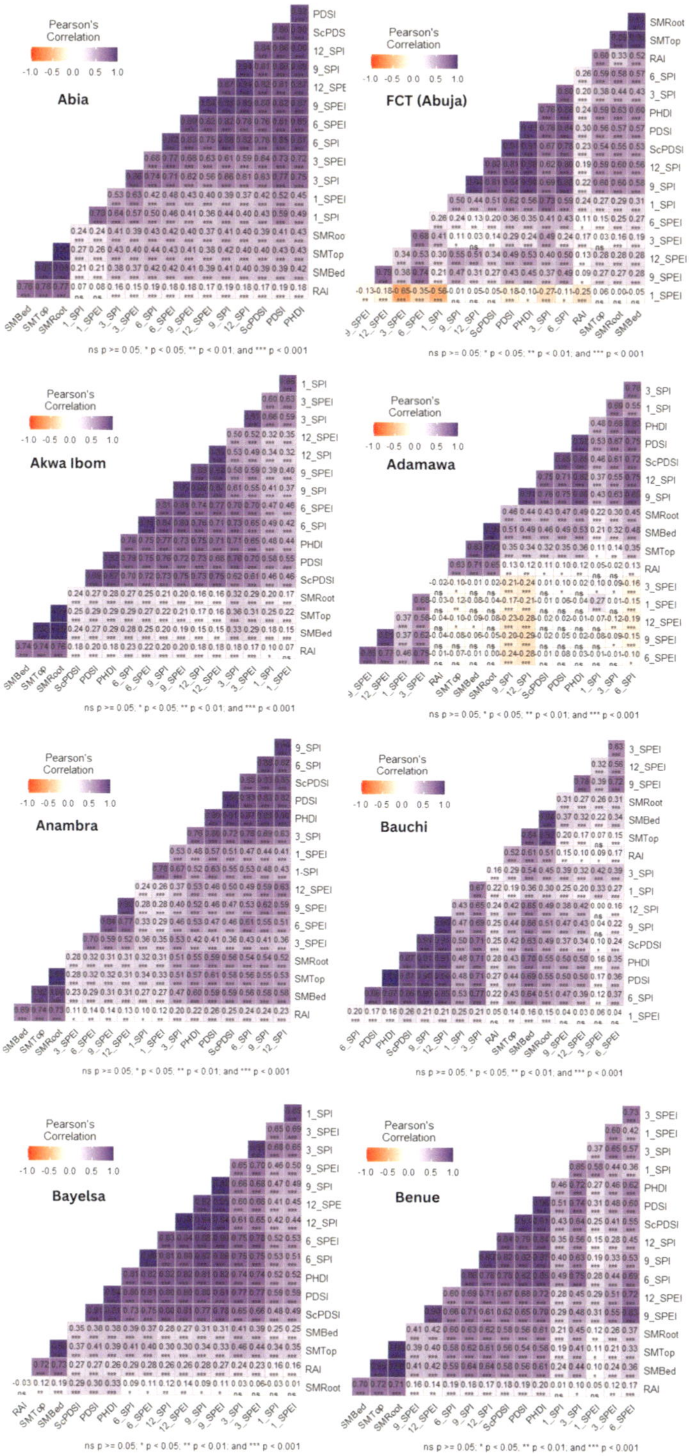
Pearson's Correlation
-1.0 -0.5 0.0 0.5 1.0
Abia
FCT (Abuja)
Akwa Ibom
Adamawa
Anambra
Bauchi
Bayelsa
Benue
ns p >= 0.05; * p < 0.05; ** p < 0.01; and *** p < 0.001

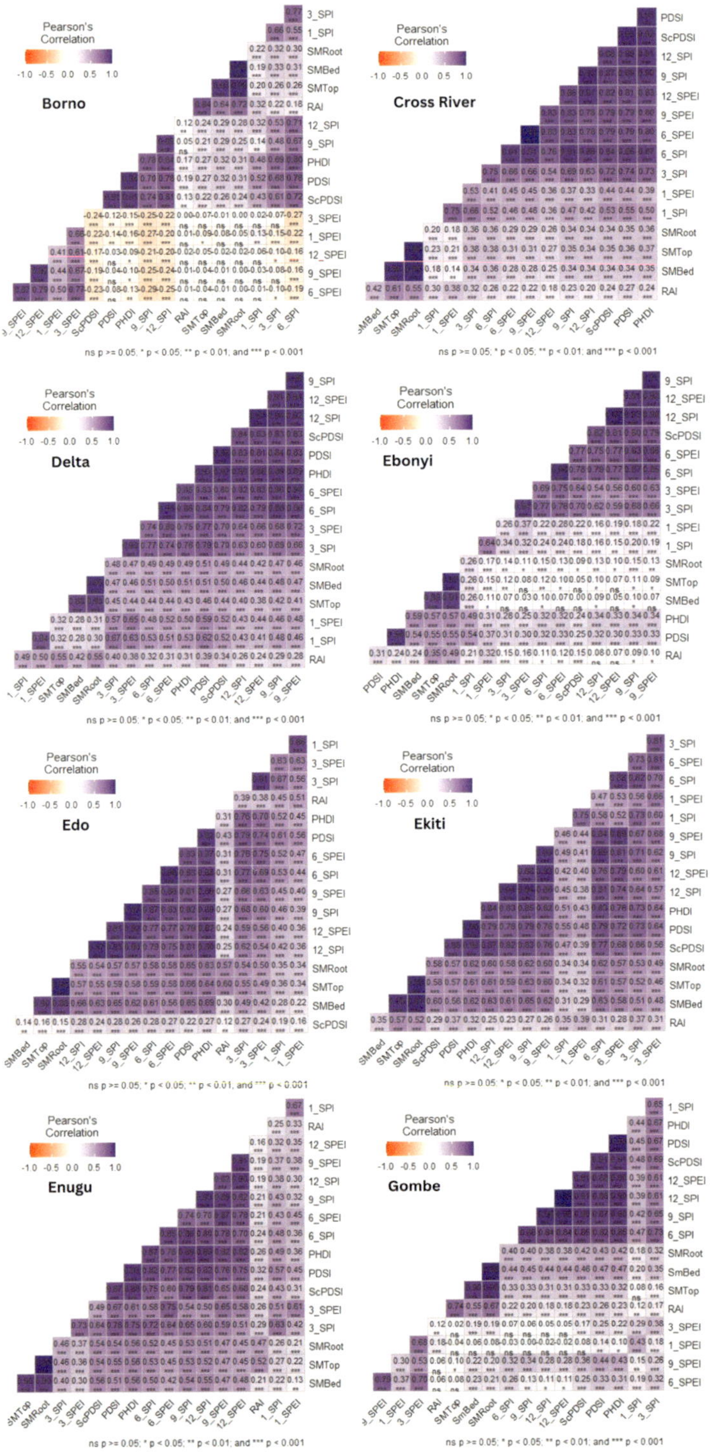
Pearson's Correlation
-1.0 -0.5 0.0 0.5 1.0
Borno
Cross River
Delta
Ebonyi
Edo
Ekiti
Enugu
Gombe
ns p >= 0.05; * p < 0.05; ** p < 0.01; and *** p < 0.001

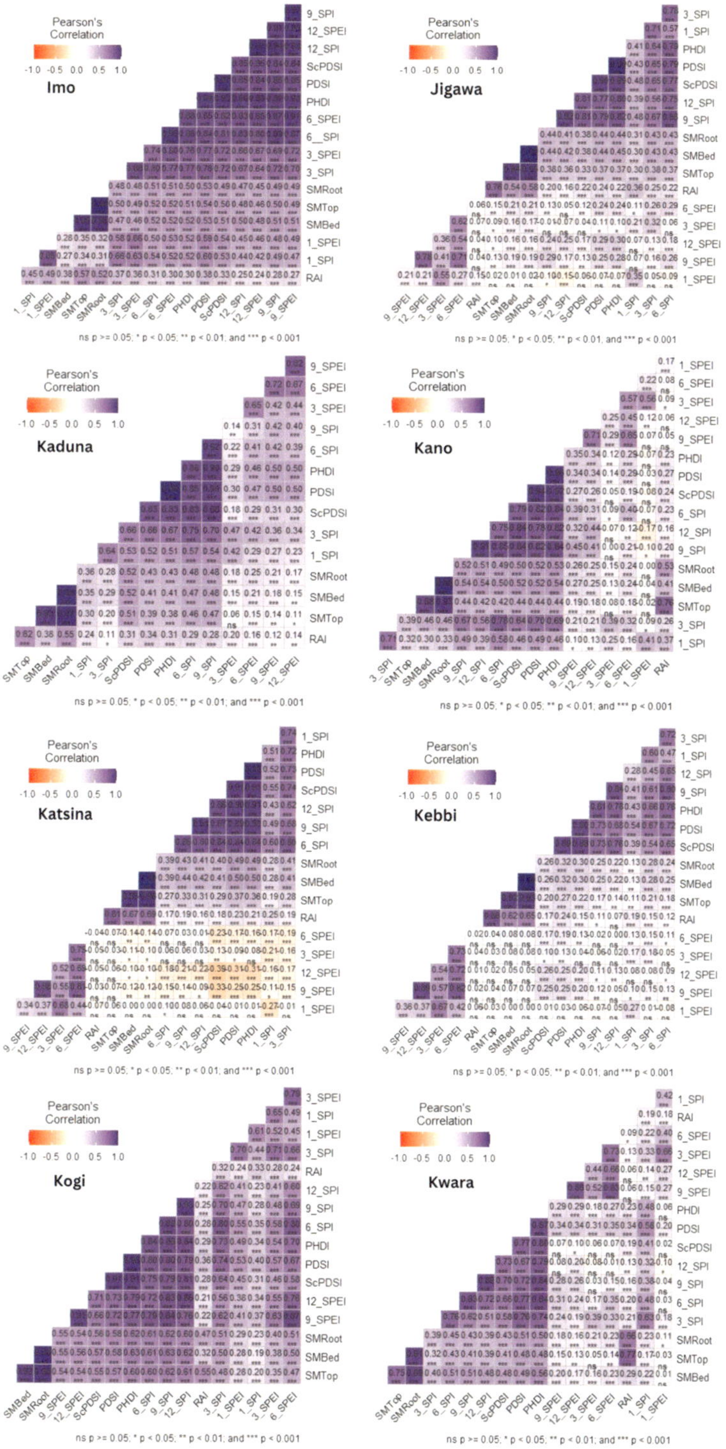
Pearson's Correlation
-1.0 -0.5 0.0 0.5 1.0
Imo
Jigawa
Kaduna
Kano
Katsina
Kebbi
Kogi
Kwara
ns p >= 0.05; * p < 0.05; ** p < 0.01; and *** p < 0.001

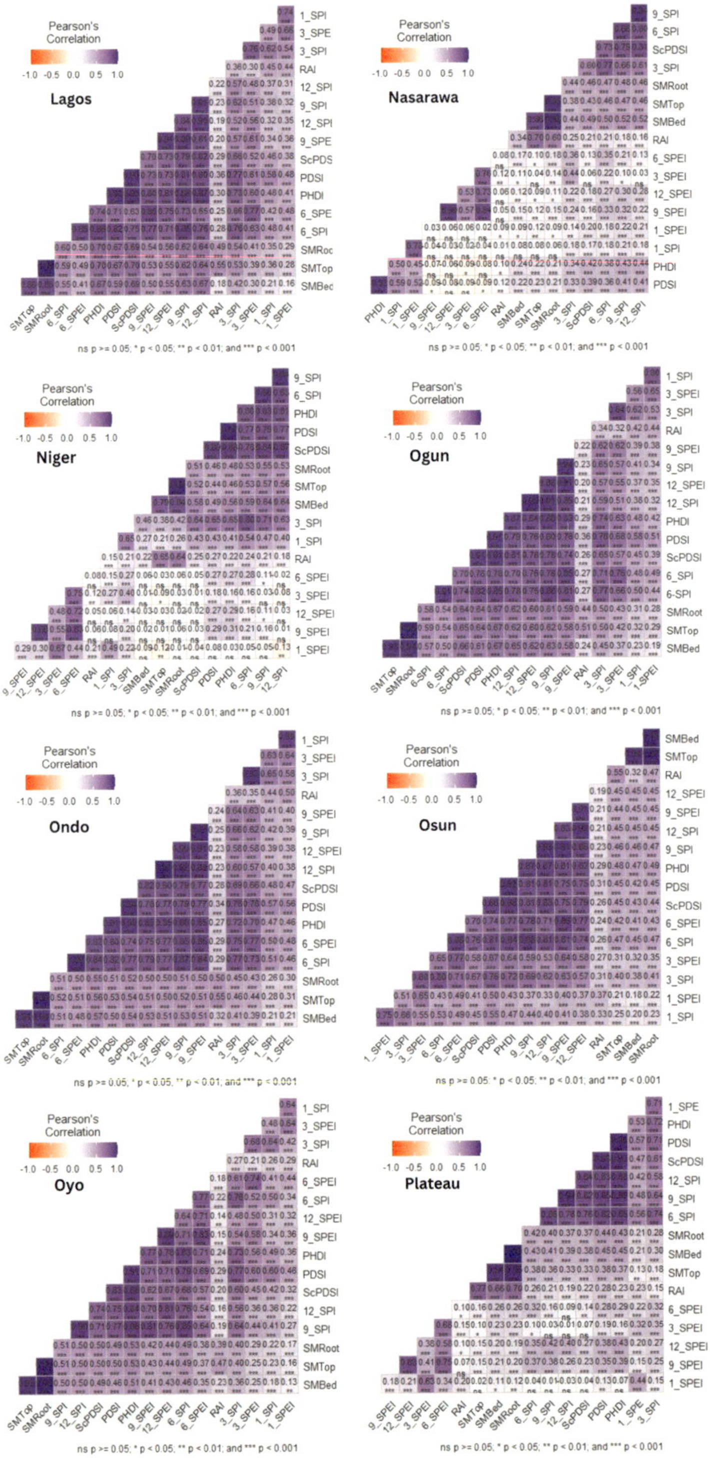
Pearson's Correlation
-1.0 -0.5 0.0 0.5 1.0
Lagos
Nasarawa
Niger
Ogun
Ondo
Osun
Oyo
Plateau
ns p >= 0.05; * p < 0.05; ** p < 0.01; and *** p < 0.001

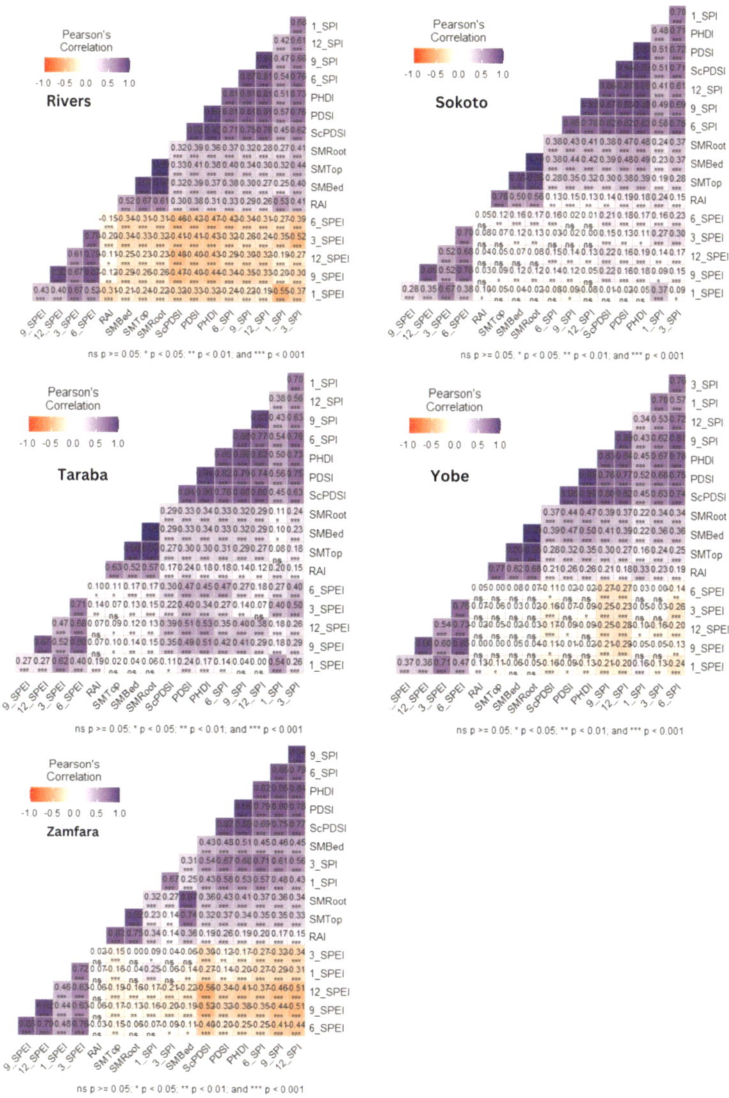

Figure 11.
Correlation of drought indices and soil moisture at various depths across all states in Nigeria.

generally all time lag variants of SPI considered in this study correlated better with soil moisture at all depth, more than all time lag variants of SPEI at all locations (**Figure 11**).

3.2 Discussion

This study investigated agricultural drought by comparing different drought indices and their variants with soil moisture at various depths across 37 locations in Nigeria. Consequently, this research provides valuable insights into the patterns, variations, and characteristics of soil moisture across Nigeria as shown in **Figure 11**. The findings of this study are of paramount importance to agriculturists, agronomists, meteorologists and engineers, among others, for the effective monitoring of soil moisture and agricultural drought conditions across the 36 states of Nigeria and the Federal Capital Territory (FCT). This would serve as part of the measure to tackle food insecurity.

Wang et al. [25] recommended the preferred use of multiscale drought indices, such as the Standardized Precipitation Index (SPI), Standardized Precipitation-Evapotranspiration Index (SPEI) and Rainfall Anomaly Index (RAI), over two-layer bucket models such as the Palmer Hydrological Drought Index (PHDI) and Palmer Drought Severity Index (PDSI). Furthermore, Wang et al. [25] suggested that deeper soil moisture is best characterized by multiscale drought indices with longer time scales. The results of this study largely align with these claims, as the monthly RAI, a multiscale drought index, correlated best with topsoil moisture storage in most regions. However, the poor performance of SPI and SPEI in correlating with monthly averages of soil moisture in many parts of this region contradicts these claims. In contrast, the two-layer bucket drought indices (PDSI, ScPDSI and PHDI), which simulate soil moisture storage in two distinct layers, outperformed most of the SPI and SPEI variants in many regions across Nigeria.

More often than not, soil moisture at the surface (SMTop) is influenced by temperature fluctuations and spatial temperature differences, particularly in open regions lacking soil or canopy cover [59, 60]. This scenario is commonly observed in arid and semi-arid regions. The soil moisture distribution pattern, which increases towards the southern region of the country, provides evidence of this phenomenon. Furthermore, the relationship between drought indices and soil moisture variation is often modulated by prevailing soil properties [61, 62]. According to Refs. [25, 61], soil properties such as bulk density and organic carbon significantly influence the relationship between soil moisture and drought indices, unlike soil texture. Predominantly, soils across Nigeria are coarse-textured, with soil texture becoming finer from north to south [63]. This implies that the northern regions are characterized by coarser sand particles than the southern region. As reported by Akpa et al. [63] the surface soil layer in Nigeria comprises sandy soil (covering 4.2×10^6 ha) and sandy loam (covering 5.3×10^7 ha). The subsoil layer is dominated by sandy clay loam (covering 5.2×10^7 ha) and clay (covering 6.9×10^6 ha).

Zhou et al. [64] characterized agricultural drought and assessed the usability of combined soil moisture products and the Soil Water Deficit Index (SWDI) on a monthly scale to establish a relationship between meteorological, agricultural, and vegetation drought in China. Through correlation analysis, the study found that the conversion of meteorological to agricultural drought and subsequently to vegetation drought does not experience a delay, particularly under drier conditions. This was further supported by the correlation between the SPI and SWDI. Furthermore, Liu et al. [65] evaluated the performance of eight drought indices in capturing soil moisture dynamics under various vegetation and land use types in China. The study reported that the Drought Severity Index (DSI) and Evaporation Stress Index (ESI) performed well in capturing soil moisture dynamics across the six main vegetation types in China. However, the effects of time lag in drought indices on soil moisture and climatic conditions (temperature and precipitation) impacted the performance of each drought index considered, with the exception of the ScPDSI and SPEI.

The results of this study revealed that the RAI exhibited a stronger correlation with surface soil moisture (SMTop) in the majority of the regions. This may be attributed to the fact that RAI is capable of capturing long-term soil moisture dynamics and serves as a more direct indicator of soil moisture, as it does not account for temperature, evapotranspiration and wind velocity [66]. The RAI was specifically designed to capture the variability of rainfall, which is the primary determinant of soil moisture variation [67–68], and is simpler to calculate compared to other tested drought indices in this study. According to Ref. [68], the trend patterns of topsoil and subsoil moistures are closely aligned with the trend patterns of rainfall in the studied regions. Simple indices that utilize only rainfall as input, such as RAI, have recorded good performances in representing the periods and intensity of drought occurrences in Nebraska [69].

In contrast to the findings of this study, Refs. [70, 71] reported a high correlation between the SPEI and soil moisture in the western/central United States and with soil moisture anomalies in Spain, respectively. Wang et al. [25] noted that the PDSI is known to have a stronger relationship with longer time scales, but performs poorly when compared to shorter time scales and shallow soil layers, where soil water can easily evaporate. However, this study found that soil moisture in certain locations in Nigeria correlated highly mostly with PHDI, a variant of PDSI. In addition, most of the high correlation coefficients reported by PHDI were at greater soil depth below the top soil surface. Therefore, PHDI appears to be more suitable for analyzing hydrological drought rather than agricultural drought. Both PDSI and PHDI can be used to estimate drought at different durations or multiple time scales, as they are based on a physical water balance model, which can aid a robust understanding of drought conditions [36, 72, 73]. PHDI offers a broader assessment of drought by considering water storage and flow in soil, groundwater and surface water bodies. PDSI, which focuses on precipitation deficits (meteorological drought), the PHDI specifically, measures water resource deficits (hydrological drought), using additional weighted factors in its calculations [42]. The positive significant correlation between PHDI and soil moisture at greater depth in many locations can thus be attributed to its suitability for the measure of hydrological droughts. Moreover, PDSI remains a widely accepted tool for monitoring droughts in many regions and locations worldwide [74, 75].

4. Conclusion

This study examines the applicability of various drought indices in monitoring and predicting soil moisture dynamics across different states in Nigeria. The findings reveal that diverse drought indices, derived from meteorological data at varying time scales, exhibit correlations with soil moisture at different layers to varying degrees across multiple locations.

The following conclusions can thus be inferred from this research:

i. Rainfall Anomaly Index (RAI) emerges as a more effective drought index for quantifying topsoil moisture (SMTop) in most Nigerian locations. This is attributed to its flexibility in estimation and the sole requirement of rainfall data, which directly impacts topsoil moisture levels, particularly in arid and semi-arid regions.

ii. As soil moisture depth increases, the applicability of RAI diminishes, while PHDI, ScPDSI, 6_SPI, 9_SPI and 12_SPI become more relevant for monitoring soil moisture below 5 cm from the soil surface. Notably, soil moisture in arid areas with minimal rainfall correlates more closely with RAI at all soil depths, whereas soil moisture in humid regions with maximum annual rainfall depth relates better with PHDI at all soil depth, thus suggesting a focus on hydrological drought rather than agricultural drought in such regions.

iii. The various SPEI and SPI variants at different time scales exhibit limited applicability in monitoring soil moisture dynamics across different layers or depths in Nigeria.

Based on these findings, the following recommendations are proposed:

i. Further research should investigate the correlation of additional soil moisture data with drought indices from more locations within each state to gain deeper insights into the behavioral patterns of the correlation.

ii. Machine learning techniques should be applied to build empirical models that examine, monitor and predict soil moisture changes from drought indices for different regions of Nigeria based on underlying patterns and relationships, enabling users to monitor agricultural drought using available meteorological data.

iii. Data on agricultural crop yields should be incorporated to examine the applicability of drought indices in assessing soil moisture shortages.

Acknowledgements

The authors are grateful to IntechOpen for their support and willingness to publish this research.

Author details

Blessing Funmbi Sasanya[1*], Akeem Abiodun Ademola[2] and
Oluwafemi Oladipupo Kolajo[3]

1 Department of Crop and Soil Science, University of Port Harcourt, Choba, Rivers State, Nigeria

2 Department of Civil Engineering, University of Ibadan, Ibadan, Nigeria

3 Department of Agricultural and Environmental Engineering, University of Ibadan, Ibadan, Nigeria

*Address all correspondence to: blessing.env@gmail.com

References

[1] Zakwan M, Ara Z. Statistical analysis of rainfall in Bihar. Sustainable Water Resources Management. 2019;**5**(4): 1781-1789. DOI: 10.1007/s40899-019-00340-3

[2] Bhuyan MDI, Islam MM, Bhuiyan MEK. A trend analysis of temperature and rainfall to predict climate change for Northwestern region of Bangladesh. American Journal ofClimate Change. 2018;**07**(02):115-134. DOI: 10.4236/ajcc.2018.72009

[3] Vicente-Serrano SM, Domínguez-Castro F, Reig F, Beguería S, Tomas-Burguera M, Latorre B, et al. A near real-time drought monitoring system for Spain using automatic weather station network. Atmospheric Research. 2022; **271**:106095. DOI: 10.1016/j. atmosres.2022.106095

[4] Wang Y, Yang J, Chen Y, Su Z, Li B, Guo H, et al. Monitoring and predicting drought based on multiple indicators in an arid area, China. Remote Sensing. 2020;**12**(14):2298. DOI: 10.3390/ rs12142298

[5] Wei W, Zhang J, Zhou L, Xie B, Zhou J, Li C. Comparative evaluation of drought indices for monitoring drought based on remote sensing data. Environmental Science and Pollution Research. 2021;**28**:20408-20425. DOI: 10.1007/s11356-020-12120-0

[6] Watson A, Miller J, Künne A, Kralisch S. Using soil-moisture drought indices to evaluate key indicators of agricultural drought in semi-arid Mediterranean Southern Africa. Science of the Total Environment. 2022;**812**: 152464. DOI: 10.1016/j. scitotenv.2021.152464

[7] Lemenkova P. Evaluating the performance of palmer Drought Severity Index (PDSI) in various vegetation regions of the Ethiopian highlands. Acta Biologica Marisiensis. 2021;**4**(2):14-31. DOI: 10.2478/abmj-2021-0010

[8] Xu ZG, Wu ZY, He H, Guo X, Zhang YL. Comparison of soil moisture at different depths for drought monitoring based on improved soil moisture anomaly percentage index. Water Science and Engineering. 2021;**14** (3):14-31. DOI: 10.1016/j. wse.2021.08.008

[9] Zheng D, Li X, Wang X, Wang Z, Wen J, van der Velde R, et al. Sam-pling depth of L-band radiometer measurements of soil moisture and freeze-thaw dynamics on the Tibetan Plateau. Remote Sensing of Environ-ment. 2019;**226**:16-25. DOI: 10.1016/j.rse.2019.03.029

[10] Fu R, Xie L, Liu T, Zheng B, Zhang Y, Hu S. A soil moisture prediction model, based on depth and water balance equation: A case study of the Xilingol league grassland. International Journal of Environmental Research and Public Health. 2023;**20**(2): 2-18. DOI: 10.3390/ijerph20021374

[11] Liu Y, Liu Y, Wang W. Inter-comparison of satellite-retrieved and global land data assimilation system-simulated soil moisture datasets for global drought analysis. Remote Sensing of Environment. 2019;**220**:1-18. DOI: 10.1016/j.rse.2018.10.0266

[12] Wang A, Shi X. A multilayer soil moisture dataset based on the gravimetric method in China and its characteristics. Journal of Hydrometeorology. 2019;**20**(8):1721-1736. DOI: 10.1175/JHM-D-19-0035.1

[13] Zhu Q, Luo Y, Xu YP, Tian Y, Yang T. Satellite soil moisture for

agricultural drought monitoring: Assessment of SMAP-derived soil water deficit index in Xiang River Basin, China. Remote Sensing. 2019;**11**(3):1-22. DOI: 10.3390/rs11030362

[14] Modanesi S, Massari C, Camici S, Brocca L, Amarnath G. Do satellite surface soil moisture observations better retain information about crop-yield variability in drought conditions? Water Resources Research. 2020;**56**(2):1-19. DOI: 10.1029/2019WR025855

[15] Xu Y, Zhu G, Wan Q, Yong L, Ma H, Sun Z, et al. Effect of terrace construction on soil moisture in rain-fed farming area of Loess Plateau. Journal of Hydrology: Regional Studies. 2021;**37**:1-11. DOI: 10.1016/j.ejrh.2021.100889

[16] Seneviratne SI, Wilhelm M, Stanelle T, Van Den Hurk B, Hagemann S, Berg A, et al. Impact of soil moisture-climate feedbacks on CMIP5 projections: First results from the GLACE-CMIP5 experiment. Geophysical Research Letters. 2013;**40**(19):5212-5217. DOI: 10.1002/grl.50956

[17] Nicolai-Shaw N, Zscheischler J, Hirschi M, Gudmundsson L, Seneviratne SI. A drought event composite analysis using satellite remote-sensing based soil moisture. Remote Sensing of Environment. 2017; **203**:216-225. DOI: 10.1016/j. rse.2017.06.014

[18] Zhou J, Wu Z, He H, Wang F, Xu Z, Wu X. Regional assimilation of in situ observed soil moisture into the VIC model considering spatial variability. Hydrological Sciences Journal. 2019;**64** (16):1982 -1996. DOI: 10.1080/ 02626667.2019.1662024

[19] Yang L, Wei W, Chen L, Jia F, Mo B. Spatial variations of shallow and deep soil moisture in the semi-arid Loess Plateau, China. Hydrology and Earth System Sciences. 2012;**16**(9):3199-3217. DOI: 10.5194/hess-16-3199-2012

[20] Yang T, Ala M, Zhang Y, Wu J, Wang A, Guan D. Characteristics of soil moisture under different vegetation coverage in Horqin Sandy Land, northern China. PLoS One. 2018;**13**(6):1-15. DOI: 10.1371/journal.pone.0198805

[21] Seneviratne SI, Corti T, Davin EL, Hirschi M, Jaeger EB, Lehner I, et al. Investigating soil moisture-climate interactions in a changing climate: A review. Earth-Science Reviews. 2010;**99**: 3-4. DOI: 10.1016/j.earscirev.2010. 02.004

[22] Teuling AJ. A hot future for European droughts. Nature Climate Change. 2018;**8**(5):364-365. DOI: 10.1038/s41558-018-0154-5

[23] Buitink J, Swank AM, Van Der Ploeg M, Smith NE, Benninga HJF, Van Der Bolt F, et al. Anatomy of the 2018 agricultural drought in the Netherlands using in situ soil moisture and satellite vegetation indices. Hydrology and Earth System Sciences. 2020;**24**(12):1-17. DOI: 10.5194/hess-24-6021-2020

[24] Palmer WC. Meteorological drought. In: US. Weather Bureau Res. Paper. Vol. 45. Washington DC: US Department of Commerce; 1965. pp. 1-58. Available from: https://www.ncdc.noaa.gov/temp-and-precip/drought/docs/palmer.pdf

[25] Wang H, Rogers JC, Munroe DK. Commonly used drought indices as indicators of soil moisture in China. Journal of Hydrometeorology. 2015;**16** (3):1397-1408. DOI: 10.1175/JHM-D-14-0076.1

[26] Querejeta JI, Ren W, Prieto I. Vertical decoupling of soil nutrients and water under climate warming reduces

plant cumulative nutrient uptake, water-use efficiency and productivity. New Phytologist. 2021;**230**(4):1378-1393

[27] Yin H, Sun, y. Characteristics of extreme temperature and precipitation in China in 2017 based on ETCCDI indices. Advances in Climate Change Research. 2018;**9**(4):218-226. DOI: 10.1016/j.accre.2019.01.001

[28] Yu L, Gao W, Shamshiri RR, Tao S, Ren Y, Zhang Y, et al. Review of research progress on soil moisture sensor technology. International Journal of Agricultural and Biological Engineering. 2021;**14**(4):32-41. DOI: 10.25165/j.ijabe.20211404.6404

[29] Kottek M, Grieser J, Beck C, Rudolf B, Rubel F. World map of the Köppen-Geiger climate classification updated. Meteorologische Zeitschrift. 2006;**15**(3):259-263. DOI: 10.1127/0941-2948/2006/0130

[30] Popoola KO, Eludoyin AO, Oladehinde GJ, Ajayi SA, Popoola OS. Rainfall and temperature variability across selected ecological regions in Nigeria. Interdisciplinary Environmental Review. 2020;**20**(3–4):311-323

[31] Shiru MS, Shahid S, Dewan A, Chung ES, Alias N, Ahmed K, et al. Projection of meteorological droughts in Nigeria during growing seasons under climate change scenarios. Scientific Reports. 2020;**10**(1):10107. DOI: 10.1038/s41598-020-67146-8

[32] Sasanya BF, Awodutire PO, Ufuoma OG. Modelling rainfall in selected agricultural hubs in Nigeria: A comparative probability distributions study. Theoretical and Applied Climatology. 2024, 2024;**155**:3599-3612. DOI: 10.1007/s00704-024-04832-x

[33] Espinoza-Dávalos GE, Arctur DK, Teng W, Maidment DR, García-Martí I, Comair G. Studying soil moisture at a national level through statistical analysis of NASA NLDAS data. Journal of Hydroinformatics. 2015;**18**(2):277-287. DOI: 10.2166/hydro.2015.23

[34] Sharma P, Singh AK, Agrawal B, Sharma A. Correlation between weather and COVID-19 pandemic in India: An empirical investigation. Journal of Public Affairs. 2020;**20**(4):1-5. DOI: 10.1002/pa.2222

[35] Wang Z, Yang Y, Zhang C, Guo H, Hou Y. Historical and future Palmer Drought Severity Index with improved hydrological modelling. Journal of Hydrology. 2022;**610**:127941. DOI: 10.1016/j.jhydrol.2022.127941

[36] Ogunrinde AT, Oguntunde PG, Olasehinde DA, Fasinmirin JT, Akinwumiju AS. Drought spatiotemporal characterization using self-calibrating Palmer Drought Severity Index in the northern region of Nigeria. Results in Engineering. 2020;**5**:1-12. DOI: 10.1016/j.rineng.2019.100088

[37] Shen S, Howard A, Yin H, Khurshed F, Akbar M. Statistical analysis of drought indices and Alberta drought monitoring. Alberta, Agriculture, Food and Rural Development, Conservation and Development Branch. 2003;**1**:7-13

[38] Zhang J, Sun F, Lai W, Lim WH, Liu W, Wang T, et al. Attributing changes in future extreme droughts based on PDSI in China. Journal of Hydrology. 2019;**573**:607-615. DOI: 10.1016/j.jhydrol.2019.03.060

[39] Adesogan SO, Sasanya BF. Water availability investigation and assessment in South - Western Nigeria. International Journal of Energy and Water Resources. 2021;**6**:457-473. DOI: 10.1007/s42108-021-00136-2

[40] Shrestha A, Rahaman MM, Kalra A, Thakur B, Lamb KW, Maheshwari P. Regional climatological drought: An assessment using high-resolution data. Hydrology. 2020;7(2): 33. DOI: 10.3390/hydrology7020033

[41] Srivastava A, Maity R. Unveiling an environmental drought index and its applicability in the perspective of drought recognition amidst climate change. Journal of Hydrology. 2023;**627**: 130462

[42] Palanisamy B, Narasimhan B, Paul S, Srinivasan R, Wangpimool W, Sith R, et al. Development and propagation of hydrologic drought from meteorological and agricultural drought in the Mekong River Basin. Hydrological Processes. 2023;**37**(7):e14935. DOI: 10.1002/hyp.14935

[43] Vicente-Serrano SM, Beguería S, López-Moreno JI. A multiscalar drought index sensitive to global warming: The standardized precipitation evapotranspiration index. Journal of Climate. 2010;**23**(7):1696-1718. DOI: 10.1175/2009JCLI2909.1

[44] Berhail S, Katipoğlu OM. Comparison of the SPI and SPEI as drought assessment tools in a semi-arid region: Case of the Wadi Mekerra basin (northwest of Algeria). Theoretical and Applied Climatology. 2023;**154**(3): 1373-1393 10.1007/s00704-023-04601-2

[45] McKee TB, Nolan J, Kleist J. The relationship of drought frequency and duration to time scales. In: Preprints, Eighth Conference on Applied Climatology. Anaheim California: American Meteor Society; 1993

[46] Achite M, Bazrafshan O, Pakdaman Z, Wałęga A, Pourhaghverdi F, Caloiero T. Uncertainty analysis of SPI and SRI calculation using bootstrap in the Mediterranean regions of Algeria. Natural Hazards. 2024;**120**:11031-11061. DOI: 10.1007/s11069-024-06642-w

[47] Merabti A, Darouich H, Paredes P, Meddi M, Pereira LS. Assessing spatial variability and trends of droughts in eastern Algeria using SPI, RDI, PDSI, and MedPDSI—A novel drought index using the FAO56 Evapotranspiration Method. Water. 2023;**15**(4):626. DOI: 10.3390/w15040626

[48] Tirivarombo S, Osupile D, Eliasson P. Drought monitoring and analysis: Standardised Precipitation Evapotranspiration Index (SPEI) and Standardised Precipitation Index (SPI). Physics and Chemistry of the Earth. 2018;**106**:1-10. DOI: 10.1016/j.pce.2018.07.001

[49] Al-Timimi YK, Osamah A. Comparative study of four meteorological drought indices in Iraq. IOSR Journal of Applied Physics. 2016; **8**(5):76-84. DOI: 10.9790/4861-0805037684

[50] Biswas B, Karmegam D. Long-term spatio-temporal analysis and trends of precipitation over semi-arid region of Rajasthan. Meteorology and Atmospheric Physics. 2023;**135**(6):53. DOI: 10.1007/s00703-023-00991-0

[51] Siddharam KJ, Basavaraja D, Nemichandrappa M, Dandekar AT. Assessment of long term Spatio-temporal variability and Standardized Anomaly Index of rainfall of Northeastern region, Karnataka, India. Climate Change. 2020;**6**(21):1-11

[52] Raziei T. Revisiting the rainfall anomaly index to serve as a simplified standardized precipitation index. Journal of Hydrology. 2021;**602**:126761. DOI: 10.1016/j.jhydrol.2021.126761

[53] Sr CCC. Demographic predictors of disciplinary sanctions among middle school victims of bullying: A quantitative correlation study. Pastoral Care in Education. 2023;(2023):1-25. DOI: 10.1080/02643944.2023.2281303

[54] Banger A, Grigolon A, Brussel M, Pfeffer K. Identifying the interrelations between subjective walkability factors and walking behaviour: A case study in Jeddah, Saudi Arabia. Transportation Research Interdisciplinary Perspectives. 2024;**24**:101025. DOI: 10.1016/j.trip.2024.101025

[55] Sims AP, Niyogi DDS, Raman S. Adopting drought indices for estimating soil moisture: A North Carolina case study. Geophysical Research Letters. 2002;**29**(8):24-21. DOI: 10.1029/2001GL013343

[56] Riley RD, Debray TP, Collins GS, Archer L, Ensor J, van Smeden M, et al. Minimum sample size for external validation of a clinical prediction model with a binary outcome. Statistics in Medicine. 2021;**40**(19):4230-4251. DOI: 10.1002/sim.9025

[57] Wang X, Ji X. Sample size estimation in clinical research: From randomized controlled trials to observational studies. Chest. 2020;**158**(1):S12-S20. DOI: 10.1016/j.chest.2020.03.010

[58] Elbeltagi A, Kumar M, Kushwaha NL, Pande CB, Ditthakit P, Vishwakarma DK, et al. Drought indicator analysis and forecasting using data driven models: Case study in Jaisalmer, India. Stochastic Environmental Research and Risk Assessment. 2023;**37**(1):113-131. DOI: 10.1007/s00477-022-02277-0

[59] Peng C, Zeng J, Chen KS, Li Z, Ma H, Zhang X, et al. Global spatiotemporal trend of satellite-based soil moisture and its influencing factors in the early 21st century. Remote Sensing of Environment. 2023;**291**:113569. DOI: 10.1016/j.rse.2023.113569

[60] Reiners P, Sobrino J, Kuenzer C. Satellite-derived land surface temperature dynamics in the context of global change—A review. Remote Sensing. 2023;**15**(7):1857. DOI: 10.3390/rs15071857

[61] Pan Y, Zhu Y, Lü H, Yagci AL, Fu X, Liu E, et al. Accuracy of agricultural drought indices and analysis of agricultural drought characteristics in China between 2000 and 2019. Agricultural Water Management. 2023;**283**:108305. DOI: 10.1016/j.agwat.2023.108305

[62] Lu XJ, Li ZB, Yan HB, Liang YJ. Spatiotemporal variations of drought and driving factors based on multiple remote sensing drought indices: A case study in karst areas of Southwest China. Journal of Mountain Science. 2023;**20**(11):3215-3232. DOI: 10.1007/s11629-023-7927-7

[63] Akpa SIC, Odeh IOA, Bishop TFA, Hartemink AE. Digital mapping of soil particle-size fractions for Nigeria. Soil Science Society of America Journal. 2014;**78**(6):1953. DOI: 10.2136/sssaj2014.05.0202

[64] Zhou K, Li J, Zhang T, Kang A. The use of combined soil moisture data to characterize agricultural drought conditions and the relationship among different drought types in China. Agricultural Water Management. 2021;**243**:106479. DOI: 10.1016/j.agwat.2020.106479

[65] Liu Q, Zhang J, Zhang H, Yao F, Bai Y, Zhang S, et al. Evaluating the performance of eight drought indices for

capturing soil moisture dynamics in various vegetation regions over China. Science of the Total Environment. 2021; **789**:147803. DOI: 10.1016/j.scitotenv.2021.147803

[66] Nguyen H, Otkin JA, Wheeler MC, Hope P, Trewin B, Pudmenzky C. Climatology and variability of the evaporative stress index and its suitability as a tool to monitor Australian drought. Journal of Hydrometeorology. 2020;**21**(10):2309-2324. DOI: 10.1175/JHM-D-20-0042.1

[67] Ge F, Xu M, Gong C, Zhang Z, Tan Q, Pan X. Land cover changes the soil moisture response to rainfall on the Loess Plateau. Hydrological Processes. 2022;**36**(11):e14714. DOI: 10.1002/hyp.14714

[68] Sasanya BF, Ademola AA, Adesogan SO. Spatiotemporal trend analysis of precipitation, temperature, and soil moisture in selected agrarian locations within Nigeria—A comparative study. Arabian Journal of Geosciences. 2024;**17**(228):1-19. DOI: 10.1007/s12517-024-12026-y

[69] Oladipo EO. A comparative performance analysis of three meteorological drought indices. International Journal of Climatology. 1985;**5**:655-664. DOI: 10.1002/joc.3370050607

[70] Vicente-Serrano SM, Beguería S, Lorenzo-Lacruz S, Camarero J, López-Moreno JJ, Azorin-Molina JI, et al. Performance of drought indices for ecological, agricultural, and hydrological applications. Earth Interactions. 2012;**16**: 1-27. DOI: 10.1175/2012EI000434.1

[71] Scaini A, Sánchez N, Vicente-Serrano SM, Martínez-Fernández J. SMOS-derived soil moisture anomalies and drought indices: A comparative analysis using in situ measurements. Hydrological Processes. 2015;**29**:373-383. DOI: 10.1002/hyp.10150

[72] Maskey S, Trambauer P. Role of hydrological modeling in drought assessment. In: Hydro-Meteorological Hazards, Risks, and Disasters. Amsterdam, Netherlands: Elsevier; 2023. pp. 277-299. DOI: 10.1016/B978-0-12-819101-9.00014-5

[73] Choukri M, Naimi M, Chikhaoui M. Drought characterization: A systematic literature review. SAINS TANAH-Journal of Soil Science and Agroclimatology. 2023;**20**(2):250-264 10.20961/stjssa.v20i2.77206

[74] Mukhawana MB, Kanyerere T, Kahler D. Review of in-situ and remote sensing-based indices and their applicability for integrated drought monitoring in South Africa. Water. 2023; **15**(2):240. DOI: 10.3390/w15020240

[75] Soylu Pekpostalci D, Tur R, Danandeh Mehr A, Vazifekhah Ghaffari MA, Dąbrowska D, Nourani V. Drought monitoring and forecasting across Turkey: A contemporary review. Sustainability. 2023;**15**(7):6080. DOI: 10.3390/su15076080